Special Electrical Machinery

Scrivener Publishing
100 Cummings Center, Suite 541J
Beverly, MA 01915-6106

Publishers at Scrivener
Martin Scrivener (martin@scrivenerpublishing.com)
Phillip Carmical (pcarmical@scrivenerpublishing.com)

Special Electrical Machinery

Edited by

Jigneshkumar P. Desai

Scrivener Publishing

WILEY

This edition first published 2024 by John Wiley & Sons, Inc., 111 River Street, Hoboken, NJ 07030, USA and Scrivener Publishing LLC, 100 Cummings Center, Suite 541J, Beverly, MA 01915, USA

For more information about Scrivener publications please visit www.scrivenerpublishing.com.

Wiley Global Headquarters
111 River Street, Hoboken, NJ 07030, USA

For details of our global editorial offices, customer services, and more information about Wiley products visit us at www.wiley.com.

Library of Congress Cataloging-in-Publication Data

ISBN 9781394193615

Front cover images supplied by Adobe Firefly
Cover design by Russell Richardson

Set in size of 11pt and Minion Pro by Manila Typesetting Company, Makati, Philippines

Printed in the USA

10 9 8 7 6 5 4 3 2 1

Contents

Preface

In the developing landscape of electrical engineering and motor technology, the demand for specialized machinery custom-made to exact application requirements has never been more dominant. This book is an endeavour to link the knowledge gap often found in the traditional discourse on motor technologies, focusing instead on a spectrum of specialized machines that are progressively pivotal in a variety of applications. From Switched Reluctance Motors to PMSM Motors and beyond, this text delves into the operational principles, applications, and control mechanisms of machines that, while critical, have received less attention in mainstream engineering literature.

The majority of texts available to students, professionals, and enthusiasts alike tend to concentrate on conventional motors and their controls, such as Induction Motors, Synchronous Motors, and Transformers, to name a few. These resources, while invaluable, often overlook the nuanced and innovative approaches being applied to specialized motors designed to meet specific operational criteria.

The book intention is not merely to introduce these machines but to offer a deep dive into their operating principles, highlighting how these principles are applied across a range of applications and how they can be harnessed through sophisticated control strategies. This approach ensures that readers are not just encountering another catalog of motor types but are being equipped with the understanding necessary to innovate, design, and optimize these machines for their unique requirements.

The subsequent chapters are meticulously organized to ensure both a seamless understanding and comprehensive coverage of the material. The journey begins in Chapter 1 with an exploration of the Brushless Direct Current (BLDC) Motor, a technology that has found widespread application in both the electric vehicle (EV) and machinery industries due to its efficiency and reliability. Following this, Chapter 2 delves deeply into the workings of the Permanent Magnet Synchronous Machine (PMSM), offering detailed insights into a technology that plays a critical role in modern electrical engineering.

In Chapter 3, attention shifts to the Hysteresis Motor, providing a thorough examination of its principles and applications. This sets the stage for Chapter 4, where the spotlight is on the Switched Reluctance Motor (SRM), identified as one of the most promising technologies in the field for its robustness and versatility.

Moving forward, Chapter 5 focuses on the Stepper Motor, highlighting its significance in the renewable energy sector, particularly in solar and wind applications, due to its precise control and reliability. Chapter 6 addresses the Universal Motor, exploring its unique ability to operate on both AC and DC power supplies, making it indispensable in a wide range of household and industrial devices.

Finally, Chapter 7 culminates the discussion by analysing torque pulsation, a critical aspect for optimizing the performance and efficiency of electric motors. This analysis not only ties together the concepts discussed in the preceding chapters but also underscores the importance of addressing this challenge in the design and application of electric motors.

This book is the zenith of extensive research, practical insights, and a deep passion for the field of motor technology.

Jigneshkumar P. Desai

1

Brushless Direct Current Motor

Jitendra G. Jamnani

Department of Electrical Engineering School of Energy Technology, Pandit Deendayal Energy University, Raisan, Gandhinagar, Gujarat, India

Abstract

The development of permanent magnet materials with high magnetic energy product lead to the development of dc machines with Permanent magnet field excitation. A BLDC motor is a motor having stator similar to a synchronous motor and permanent magnets in the rotor, operates in self-controlled mode by the use of position sensors to detect rotor poles and an inverter for controlling the phase currents of stator winding.

In conventional dc motor, the commutator and brushes are needed, which are subjected to wear and require maintenance. In BLDC motor, the function of commutator and brushes are implemented by using semiconductor power switches and digital controllers and hence maintenance free motor can be released. This motor is called BLDC motor. The rotor position sensors and the semiconductor power switches in the inverter performs the role of the commutator and brushes of the conventional dc motor. A conventional dc motor has stationary field system which consists of field magnets. It produces magnetic field in the machine. The armature is rotating part and it is the rotor. The armature is free to rotate between the field poles and it is mounted on the shaft. The construction of BLDC motor is similar to the AC motor called Permanent Magnet Synchronous Motor (PMSM).

The brushless dc motors offer many advantages as compared to conventional dc motors. BLDC motors have small size rotor, high power density, low inertia, higher efficiency and requires lower maintenance. High speed and torque capability because of absence of commutator and brushes.

Presently BLDC motors are the significant contributors for the modern drive technology. They are used for a wide variety of applications in the fields of Electric vehicles, hybrid vehicles, robotics, computer disk drives, servo drives, DVD players, fans, washing machines, pumps, blowers, compressors, Industrial robots and CNC machine tools etc.

Email: jg.jamnani@sot.pdpu.ac.in

Jigneshkumar P. Desai (ed.) Special Electrical Machinery, (1–14) © 2024 Scrivener Publishing LLC

Keywords: Brushless DC motors, position sensors, permanent magnet, permanent magnet synchronous motor (PMSM), BLDC drive

1.1 Brushless DC (BLDC) Motors

A BLDC motor is a motor having stator similar to a synchronous motor and permanent magnets in the rotor, operating in self-controlled mode by the use of position sensors to detect rotor poles and an inverter for controlling the phase currents of stator winding.

In conventional dc motor, the commutator and brushes are needed, which are subjected to wear and require maintenance. In BLDC motor, the function of commutator and brushes are implemented by using semiconductor power switches and digital controllers and hence maintenance free motor can be released. This motor is called BLDC motor. The rotor position sensors and the semiconductor power switches in the inverter performs the role of the commutator and brushes of the conventional dc motor [1].

1.2 Construction of Brushless DC (BLDC) Motors

The construction of BLDC Motor is shown in Figure 1.1. A BLDC Motor has two main parts: A stator- stationary part and a rotor – a rotating part. The stator consists of stator core and 3-phase AC distributed winding. The windings are similar to those in synchronous motor.

The construction of modern BLDC motor is similar to the AC motor called PMSM. The construction of PMSM is the same as conventional

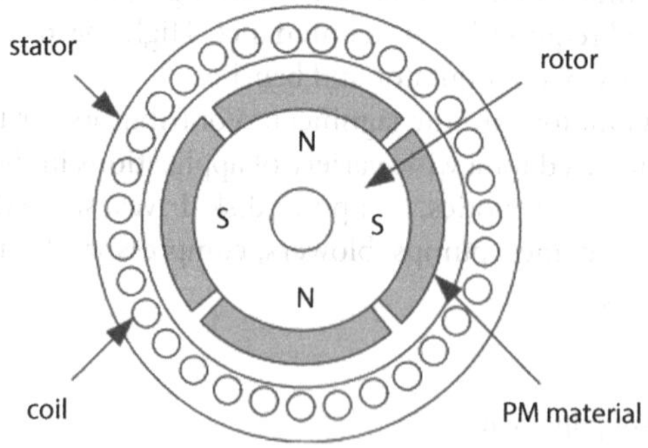

Figure 1.1 Construction of BLDC motor.

synchronous motor, but the only difference is with the rotor. The rotor consists of permanent magnets to create field poles instead of wound field winding.

Instead of wound field for the rotor, permanent magnets are mounted for creating rotating magnetic field. Since no dc supply is needed for exciting the rotor poles, these motors are very simple, robust, reliable and low cost.

The PMSM are very reliable, brushless, robust and gives very fast response when compared to the conventional motors. It produces ripple free torque, lower noise and suitable for high-speed applications.

1.3 Brushless DC Motor Drive System

The main parts of a BLDC drive system are shown in the schematic block diagram of Figure 1.2. The drive motor has three elements: stator, a rotor carrying a permanent magnet excitation system and a non-contacting means of sensing rotor position (rotor position sensors).

For a 3 phase motor supplied from a six-step power conditioner, the operation of the system is as follows:

The control circuit receives information on the position of rotor from the position sensors. This is then translated into one of six current states as shown in Figure 1.4. Each current pulse produces an EMF in the stator which remain fixed in space for 60^0 electrical until the next current state. Once the rotor completes 60° electrical rotation, the phase currents are ideally switched instantaneously into the next state. This results in a step of armature emf of 60° in the direction of rotation. This process will be repeated six times per electrical revolution resulting in a discretely emf and continuous rotation.

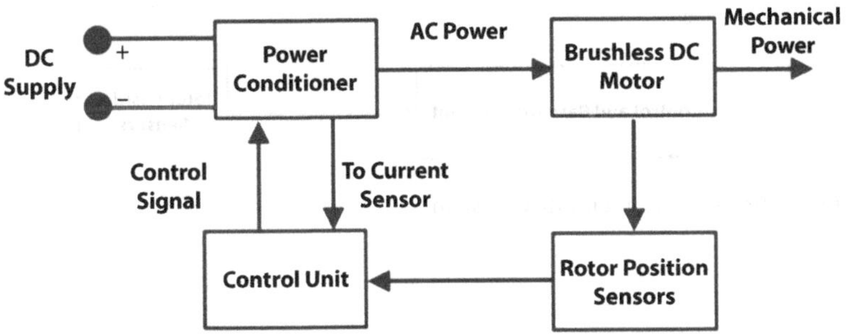

Figure 1.2 Schematic block diagram of brushless DC drive system.

The BLDC motor operates in the same way as self-controlled synchronous motor. The semiconductor power switches in the inverter controls the motor current are commutated by the back emf of the motor.

Nowadays, the inverter which consists of semiconductor power switches such as IGBTs or MOSFETs are used. The ON and OFF states of power switches are controlled by the gate signals, that can be obtained from rotor position sensors.

Figure 1.3 shows an inverter configuration which is fed from dc source and used for a BLDC drive system. The turn-on and off instants for the power swithes are controlled by rotor position sensors such that the angle between the rotor and stator field is regulated at 90°. The waveforms of the stator phasecurrents are shown in Figure 1.4.

On every 60^0 electrical the inverter switches will be turned on, the power switches are given numbers in the sequence in which they are turned ON. If the inverter is suppled from a DC voltage (VSI), the PWM of the individual switches can provide the control of the motor phase current [2].

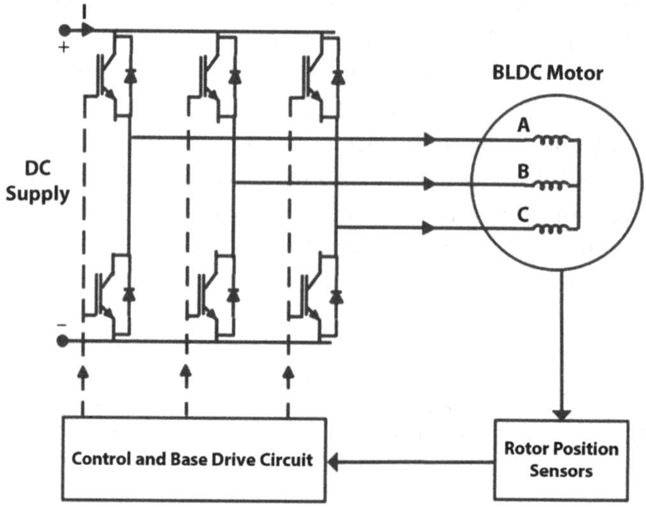

Figure 1.3 Brushless dc motor drive system.

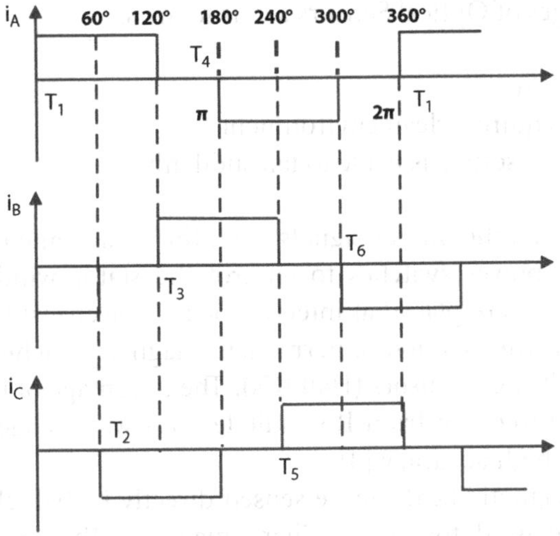

Figure 1.4 Idealised stator phase currents.

1.4 Position Sensors

The position sensors can be used for detecting the position of the rotor poles and send logic codes to a commutation decoder and thereafter processing this code, the firing circuits of semiconductor power switches will activate for feeding the power to the stator winding of the driving motor. A unidirectional torque is produced due to the interaction between the permanent magnets of the rotor and currents flowing through the stator winding. A BLDC motor incorporating rotor position sensors [3]. Nowadays the following position sensors are used:

(a) Optical Sensors
(b) Hall Sensors or Magnetic field sensors

In case of optical sensors, a light source shines through a patterned disc which is attached to the shaft of rotor and a photodiode will detect the presence or absence of light. Both the light source and photodiode are stationary. The main advantage of these sensors is that the signal from the photodiode rises and falls quite abruptly and hence switching points are well defined. Also, the output signal is dc, so it does not require rectification or filtering.

Disadvantages of Optical Sensors are as follows:

1) High cost
2) They require a clean environment
3) The light source is liable to fail suddenly

In Hall sensors, the output signals from the Hall elements operate the semiconductor power switches to control the stator winding currents. These hall sensors are placed at intervals of 120° electrical to detect rotor magnetic poles. The position of permanent magnets attached to the rotor is sensed by hall effect sensors (Hall ICs). The advantage of magnetic field hall sensors are freedom from RFI, suitale for a wide range of operating conditions and high accuracy [4].

The rotor magnetic field can be sensed directly by hall elements (ICs) eliminating the need for any auxiliary magnets. The sensors must be located far enough from the stator winding to prevent the magnetic field generated by stator currents from interfering with their operation [5].

1.5 Features and Advantages of BLDC Motors

BLDC motors have the following advantages and special features as compared with conventional brushed dc motors and Synchronous motors:

- Due to the absence of commutator, brushes and field windings, the size of the rotor is small and the motor has high power density.
- The inertia is low and hence dynamic response is fast
- Due to the absence of wound field, No field copper loss and hence efficiency increases
- Elimination of brushes and slip rings, no sparking
- Requires less maintenance and lower maintenance cost
- The motor has very high speed and torque capability because of absence of commutator and brushes
- High torque to inertia ratio
- As the armature winding is stationary, the heat dissipation is better
- High efficiency and reliability
- Torque ripples are absent
- Produces low noise

1.6 Permanent Magnet Rotor Configuration

The following are the commonly used permanent magnet rotor configurations:

(i) Surface-mounted
(ii) Interior or Buried Magnets

Figure 1.5 shows the 4-pole surface mounted PMSM and Figure 1.6 shows 6-pole Interior or Buried magnet PMSM [6] . In surface mounted type, the permanent magnets projects from the surface of the rotor. In interior type, the magnets are embedded in the interior of the rotor. Magnets are inserted into the rotor and hence provides smooth rotor surface [10].

Due to the constant gap between the stator and rotor, the surface mounted magnets can provide square wave flux distribution. Hence surface mounted magnets are used in trapezoidal type BLDC motors.

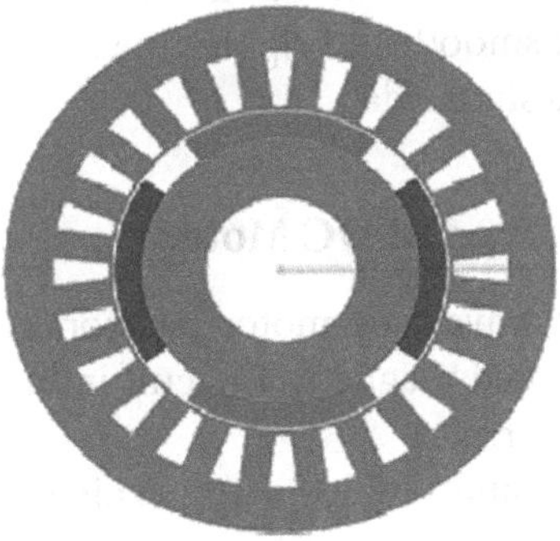

Figure 1.5 Four-pole surface mounted PMSM.

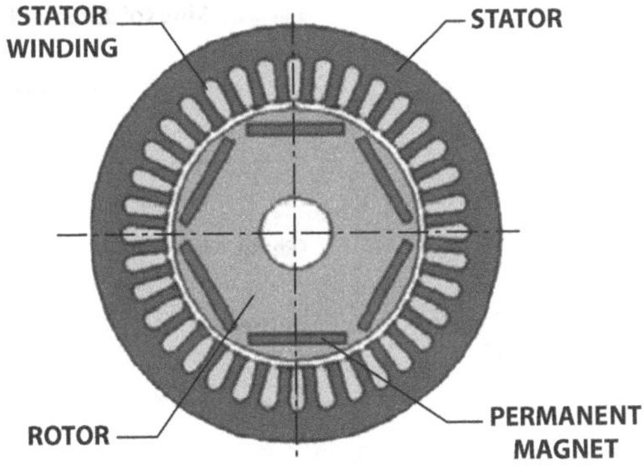

Figure 1.6 Six-pole interior or buried magnet PMSM.

Whereas by proper design of rotor geometry, the interior or buried magnets can provide sinusoidal flux distribution and hence the interior magnets are used for sinusoidal type BLDC.

1.7 Types of BLDC Motors

Depending up on the nature of voltage induced in the stator the BLDC Motors are classified as follows:

(i) Trapezoidal or Square wave type and
(ii) Sinusoidal type

In the trapezoidal type BLDC motor, the nature or shape of the induced voltage is of trapezoidal and in order to have smooth and ripple free torque operation, the phase current requirement is 120° quasi square wave.

In the sinusoidal type BLDC motor, the shape of Induced voltage will be sinusoidal and to have smooth and ripple-free torque operation, the phase current requirement is sinusoidal.

1.7.1 Trapezoidal Type BLDC Motor

Trapezoidal or square wave type motor has concentrated windings. The rotor has surface mounted magnets. The emf induced in the stator phase has a trapezoidal waveform.

Figure 1.7 shows the sinusoidal and Trapezoidal waveform generated in BLDC motor [7].

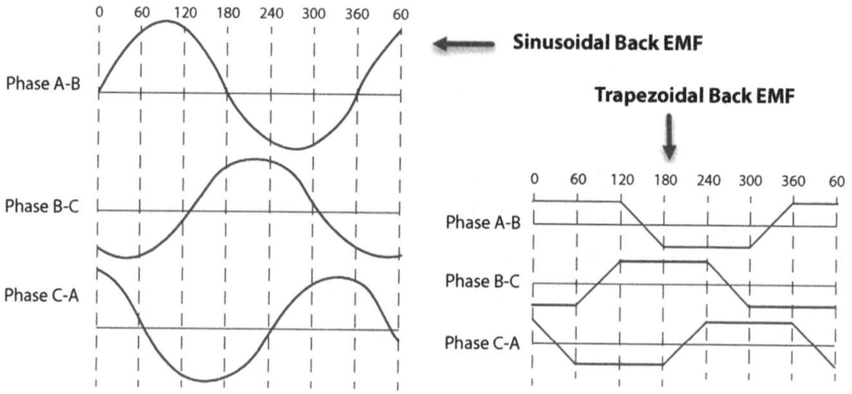

Figure 1.7 Sinusoidal and trapezoidal back emf wave forms of BLDC motor [7].

The advantages and disadvantages of Trapezoidal type BLDC Motors are as follows [8]:

1.7.1(a) Advantages

- It need an inexpensive position sensors
- Its Control scheme is simple
- High inverter efficiency as only two switches will be ON at any time
- Overall cost will be less
- Less switching losses

1.7.1(b) Disadvantages

- Torque ripple at every commutation, torque ripple will be high
- Torque produced is less
- Acoustic and electric noise

1.7.2 Sinusoidal Type BLDC Motor

In Sinusoidal type BLDC motor, the stator has distributed winding, similar to synchronous motor. The rotor consists of inset or interior type permanent magnets. Rotor poles are so shaped that the emf induced in a stator phase will have sinusoidal waveform [9].

Sinusoidal type BLDC Motor has the following advantages and disadvantages:

1.7.2(a) Advantages

- Torque ripple will be lower
- Smooth motion
- Maximum Torque is produced

1.7.2(b) Disadvantages

- It requires an expensive position sensor
- Requires more hardware and signal processing
- Inverter efficiency will be low because 3 switches are on at the same time.
- Overall cost will be higher

1.8 Square Wave Brushless Motor

Figure 1.8 shows the square wave Brushless motor. Here, the two-pole magnet has a pole arc of 180°. The waveform of the airgap flux density distribution is a square wave. With 180° magnet arcs and 2 slots/pole/phase, the flat top of the waveform will be ideally 150° wide but reduces to 120° due to fringing effect.

The stator has 3-phase ac distributed winding and the number of slots are 12. Thus there are 2 slots/pole/phase. Each phase winding contains two adjacent full pitch coils of each of N_1 turns, the axes are displaced from each other by 30° degrees. The winding is single layer and each slot contains N_1 conductors from one phase winding.

The Induced EMF in coil a_1A_1 is given by Equation 1.1,

$$e_1 = -\frac{d\psi_1}{dt} = -\omega\frac{d\psi_1}{dt} \tag{1.1}$$

Where, $e_1 = 2\,N_1B_g\,l\,r_1\omega$ Volt, $\psi_{1max} = N_1B_g\pi\,l\,r_1, \phi = B_g\pi l r_1 = B_gA$, A = area of flux path, l = axial length of magnet, r_1 = radius, e_1 = indicates the magnitude of the square wave e.m.f. e_{a1}.

The induced EMF in the second coil of phase A will be identical, but retarded in phase by 30°, shown in Figure 1.9.

The resultant phase voltage will be the sum of two separate coil voltages as the two coils are connected in series. The effect of distributed winding into two coils is to produce the waveform of stepped EMF.

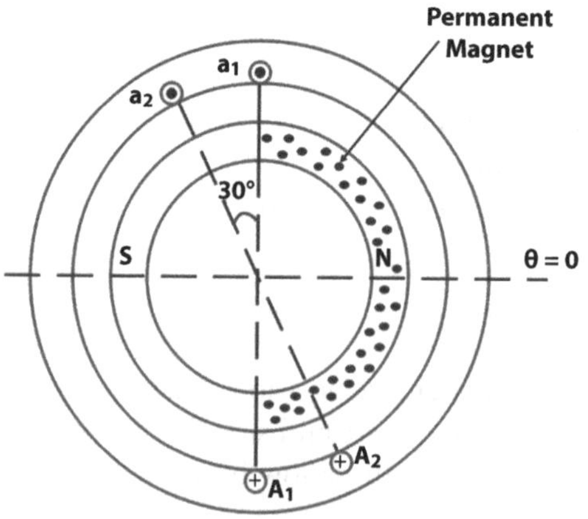

Figure 1.8 Square wave brushless motor: Motor showing two coils of one phase

The waveform has a trapezoidal shape that is characteristic of the brushless dc motor as shown in Figure 1.9. (Fringing causes its corners to be rounded, indicated as dotted lines).

With 180° magnet arcs and 2 slots/pole/phase, the flat top of the waveform will be 150° wide, but due to fringing effect reduces to 120°. The magnitude of the phase emf (flat-topped) phase will be given as Equation 1.2) below,

$$e = 2\,N_{ph} B_g\, l\, r_1 \omega \text{ Volt} \tag{1.2}$$

Where, N_{ph} = No. of turns in series per phase, ω = mechanical radian/ mechanical radian/seconds

In this case, $N_{ph} = 2\,N_1$, considering the two coils connected in series

Mechanical power developed (2 indicates two phases are conducting) is given by Equation 1.3 below,

$$P = \omega T_e = 2eI \tag{1.3}$$

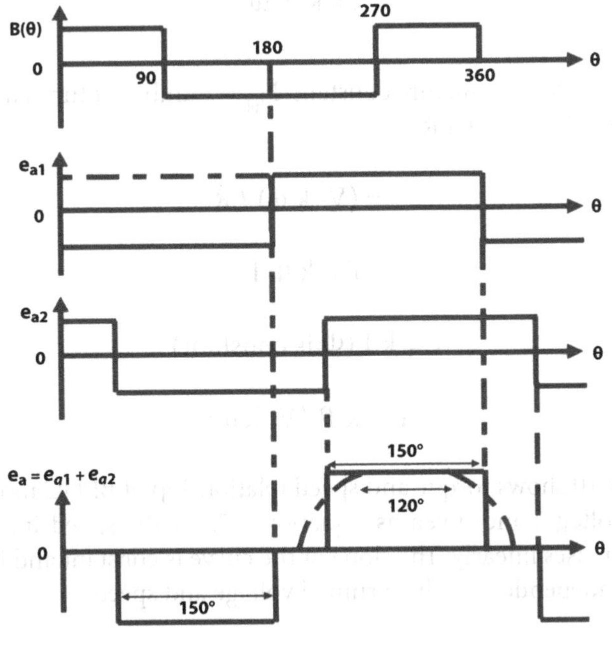

Figure 1.9 Waveforms of square wave BLDC motor.

Electromagnetic Torque is given by Equation 1.4 below,

$$T_e = \frac{2eI}{\omega}$$

$$T_e = 4N_{ph}B_g\, l\, r_1 I \quad \text{N-m} \tag{1.4}$$

This is the equation of Electromagnetic torque in square wave BLDC motor.

1.9 Torque/Speed Characteristics of BLDC Motor

In BLDC motor, the voltage equation is given as Equation 1.5,

$$V = e + I \times r \tag{1.5}$$

Where, r = summation of resistances of two phases in series, e = summation of two phase voltages in series. In which e is given as Equation 1.6,

$$e = k\, \Phi\, \omega \tag{1.6}$$

where, k = 4 , N_{ph} = armature constant, N_{ph} = number of turns in series

Therefore, V = k ω + I R

$$I = (V - k\,\omega)\,/\,R$$

$$T = k\,\Phi\, I$$

$$T = k\, I\ (\Phi \text{ is constant})$$

$$T = k/R\,(V - k\,\omega) \tag{1.7}$$

Figure 1.10 shows torque and speed relationship of BLDC motor at two different voltages and given as Equation 1.7. As the speed increases the torque decreases linearly. The slope of the curve is constant and is equal to k^2/R. It is independent of the terminal voltage and speed.

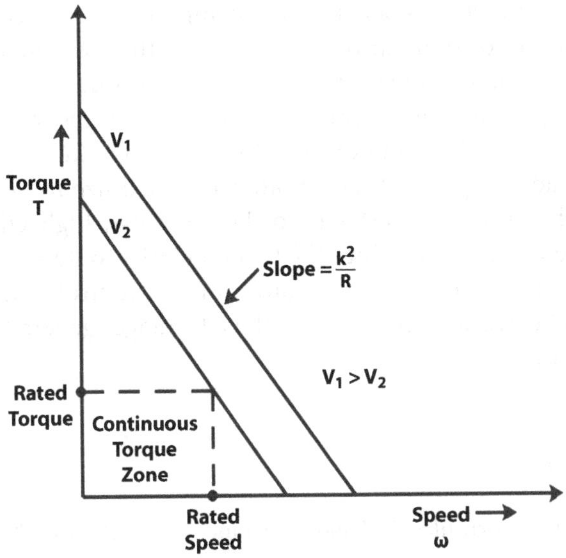

Figure 1.10 Torque versus speed curves of BLDC motor.

1.10 Applications of BLDC Motors

Brushless DC motors (BLDC) are widely used for the application in the fields of industrial control, automotive, aviation, automation systems etc [10]. The commonly used applications of BLDC motors include:

- Disc drives
- DVD and CD players
- Electric Vehicles and other battery-powered applications
- Servo drives
- Robotics applications, machine tools, and conveyors
- Washing machines
- Compressors, dryers and fans
- Pumps and blowers

1.11 Conclusion

This chapter deals with the construction of BLDC motor, BLDC motor drive system, Rotor configurations commonly used for PMBLDC motors.

The types of BLDC motors and their comparison have been discussed. The type of rotor position sensors used to detect the position of rotor poles are also discussed. The square wave type brushless dc motor along with EMF and torque equations is presented. The advantages, various features and applications of BLDC motors are also included in the chapter.

The main advantages of BLDC motors are small size rotor, high power density, High torque to inertia ratio, Lower noise, high efficiency and requires little maintenance. The BLDC motor drive system requires rotor position sensors, power electronic convertor and control unit resulting in higher cost of drive system. However the advantages generally overweigh the limitations.

References

1. T. J. E. Miller, "Brushless PM and Reluctance Motor Drives", Clarendon Press Oxford
2. R. Krishnan, "Electric Motor Drives", PHI
3. Jacek Gierasewing, "P. M. motor technology", Marcel Dekker Publications
4. Denis O'Kelly , "Performance and Control of Electrical Machines", McGraw-Hill, 1991
5. P.C. Sen, "Principles of Electrical Machines and Power Electronics", John Wiley & Sons.
6. T. Kenjo and S. Nagamori, "Permanent magnet and Brushless dc motors", Clarendon press, Oxford
7. Abu Hassan, Mohd Arif, Abdul Rahim Abdullah, Nurhazilina Bahari, and Muhammad Ihsan Mohd Sabri. "Efficiency comparison of trapezoidal and sinusoidal method for brushless dc motor drive." In Applied Mechanics and Materials, vol. 785, pp. 248-252. Trans Tech Publications Ltd, 2015
8. Miyamasu, Masataka, and Kan Akatsu. "Efficiency comparison between Brushless dc motor and Brushless AC motor considering driving method and machine design." IEEJ Journal of Industry Applications 2, no. 1 (2013): 79-86.
9. Niapour, S. A., G. H. Garjan, Mehdi Shafiei, Mohammad Reza Feyzi, Saeed Danyali, and Mojtaba Bahrami Kouhshahi. "Review of permanent-magnet brushless DC motor basic drives based on analysis and simulation study." International Review of Electrical Engineering 9, no. 5 (2014): 930-957.
10. Muhammed Fasil, Nenad Mijatovic, Bogi Bech Jensen, and Joachim Holboll, "Performance Variation of Ferrite Magnet PMBLDC Motor With Temperature, IEEE Transactions on Magnetics, Vol. 51, No. 12, December 2015.

2

Permanent Magnet Synchronous Motors

Mousumi Jana Bala[1]* and Chandan Jana[2]

[1]*Department of Electrical Engineering, Meghnad Saha Institute of Technology, West Bengal, India*

[2]*Department of Electrical Engineering, Hooghly Engineering and Technology College, West Bengal, India*

Abstract

Permanent Magnet Synchronous Motors (PMSM) are an evolved form of conventional synchronous motors, distinguished by the replacement of the field electromagnet with a permanent magnet. This alteration enhances efficiency by 10% compared to traditional synchronous motors, attributed to reduced field copper loss and heightened torque and power density.

PMSMs find extensive applications in automation, agriculture, military, aerospace, medical, and electric vehicles, with global markets spanning North and South America, Europe, the Middle East, Africa, and Asia Pacific, particularly China. Key manufacturers include General Electric, DRS Technology, Rockwell Automation, Mitsubishi Electric, Siemens, Toshiba, Emerson, ABB, WOLONG, and WEG, offering power ratings from several kW to 600 kW.

Historically, Ferrites and Alnico magnets were used in PMSMs but faced challenges like brittleness and low remanence. The breakthrough in the 1980s came with rare-earth permanent magnets like Samarium-Cobalt and Neodymium-Iron-Boron, overcoming demagnetization issues. PMSMs are categorized as radial or axial flux motors, and by induced EMF nature, into sinusoidal or trapezoidal types. The rotor's permanent magnet placement further distinguishes them into Surface Mount PM, Inset PM, and Interior PM synchronous motors.

The chapter explores permanent magnet properties, PMSM classifications, operating principles, steady-state circuits, dynamic models, and control strategies in five sub-topics, providing a comprehensive understanding of Permanent Magnet Synchronous Motor technology.

**Corresponding author*: mousumibala@rediffmail.com

Jigneshkumar P. Desai (ed.) Special Electrical Machinery, (15–60) © 2024 Scrivener Publishing LLC

Keywords: Permanent magnet, demagnetisation, halbach array, surface mount PMSM, inset PMSM, cogging torque, vector control, field weakening

2.1 Fundamentals of Permanent Magnets

Permanent Magnet produces the magnetic field in the air gap without consumption of any electrical energy. PMs have a wider hysteresis loop and hence called as *hard ferromagnetic materials*. The portion of the B-h curve in the second quadrant is known as the *demagnetisation curve* and the characteristics shown by the PMs in this region are called the *demagnetisation characteristics*. The parameters describing the PMs Characteristics are discussed below:

2.1.a Remanence or *remanent magnetic flux density,* B_r: The value of the magnetic flux density at zero magnetic field intensity is known as remanence or remanent flux density, $\mathbf{B_r}$. For maintaining higher value of magnetic flux density the permanent magnet should have high value of remanence. The SI unit of remanence is Tesla (T) where $1T=1Wb/m^2$ in MKS unit or Gauss where 1 Gauss= 0.0001T

2.1.b Coercivity *or coercive field intensity,* $\mathbf{H_c}$**:** The value of the demagnetizing field intensity required to bring the magnetic flux density of a previously magnetized material to zero, is known as coercivity, H_c. A magnet with high value of coercivity requires a thinner volume to withstand the demagnetisation field. The SI unit of coercivity is A/m or Oersted (Oe) where 1 A/m = $4x10^{-3}$ Oe. Figure 2.1 shows B_r, H_c on the B-H curve and demagnetisation characteristics.

2.1.c Energy product and *maximum energy product,* $\mathbf{(BH)_{max}}$**:** The energy product of a PM is defined as the absolute value of the product of the magnetic flux density, B and field intensity H at each point of the demagnetisation curve of that particular PM. The maximum energy product is the maximum value of the energy product in the second quadrant. This value is one of the main parameters of PM which requires for calculation of the dimension of PM in the electrical machine. The SI unit of maximum energy product is J/m^3 or kJ/m^3. Figure 2.2 shows the energy product as a function of flux density B [1].

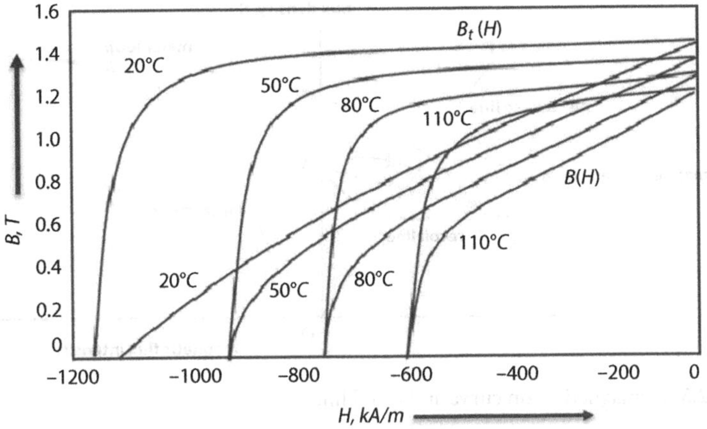

Figure 2.1 B-H curve and Demagnetisation characteristics.

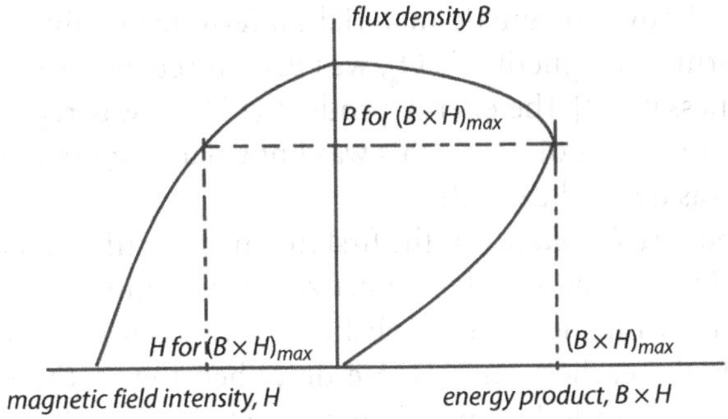

Figure 2.2 Energy product and demagnetisation curve.

2.1.d Recoil line and Recoil permeability, μ_{rec}: For a previously magnetized material application of negative field intensity will reduce the magnetic flux density at a point say A, Withdrawal of the negative field will increase the flux density to point C which is slightly less than the remanence following a minor hysteresis loop. Reapplying the negative field again will reduce the magnetic flux density to the point A again, according to the minor hysteresis loop. This minor hysteresis loop may be approximated by a straight line called recoil line. The line AC in Figure 2.3 is the recoil line. The slope of the recoil line is known as the recoil permeability, μ_{rec}.

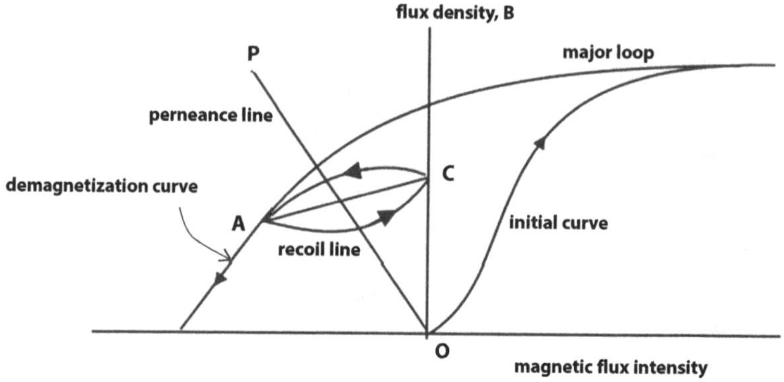

Figure 2.3 Demagnetisation curve and recoil line.

2.2 Early History of Permanent Magnet

The first man known magnetic material ***Lodestone*** or mineral magnetics which is a form of magnetite Fe_3O_4, was discovered as early as BC 600 in Magnesia, Thessaly [2].The energy product $(BH)_{max}$ was reported as 1 kJ/m^3. After their place of discover, this was known as ***Magnets*** and the word ***Magnetism*** was derived thereafter.

As reported, ***needle magnet*** is the first manmade artificial magnet which was nothing but an iron needle magnetized by touching the loadstone. In 1660,a report on magnets was published by William Gilbert which was considered as the earliest one [3]. He described the method to increase the attractive force of lodestone when associated with soft iron pole tips. Invention of the electromagnet in 1825 by J. Henry and W. Sturgeon lead the advancement a few step forward[4].

In 1901, ***Heusler Alloys*** [5] (CuMnAl) were reported having outstanding properties compare to the existing magnets. Heusler alloy is a composition of 10-30% Manganese, 15-19% Aluminium and Copper as the rest part. Japanese physicist K. Honda and T. Taki discovered ***Cobalt steel alloy*** in 1917 with a $(BH)_{max}$ of 7.6 kJ/m^3 [6]. Next great leap in the advancement of the permanent magnet development is the manufacturing of ***Al-Ni-Fe*** magnet in 1930 with a $(BH)_{max}$ as high as 10 kJ/m^{3}and a high value of coercivity. Addition of ***Cobalt*** (Co) with ***Al-Ni-Fe alloy*** in 1931, give the birth of ***Alnico*** magnet in Japan. The energy product of Alnico magnet could be as high as 36 kJ/m^3. The development of ***powder magnets*** in 1938, again in Japan by Kato and Takei was the portent of the modern ferrite. Now a days, permanent magnets with a $(BH)_{max}$ as high as 415 kJ/m^3 can be obtained [7].

The modern permanent magnets are classified as the follows:

- Alnico magnet
- Ceramic magnet
- Rare earth permanent magnet

2.2.1 Alnico Magnet

As stated earlier alnico magnets are Al-Ni-Co-Fe alloy. Alnico magnet provides a very high remanence flux density and low temperature coefficient of B_r near about –0.02% per °C with a maximum service temperature of 520°C. But at the same time it has a lower value of coercive force and its demagnetisation curve being non-linear in nature it could be easily demagnetized. Two opposite poles placed near to each other can weaken themselves. That's why Alnico magnets are normally magnetized length wise and sometimes is used with additional mild steel pole shoes. Alnico magnets dominates the market from mid 1940s up to the development of ceramic magnets in 1960s.

2.2.2 Ceramic Magnet

Ceramic magnets which are also known as Ferrite magnets have lower value of remanence flux density but a higher value of coercive force than Alnico. Because of high coercive force, it is possible to magnetize it across width wise. Temperature coefficients are also relatively high as much as –0.2% per °C and maximum service temperature is 400°C. Due to its lower cost than Alnico, ferrite magnets are most economical in fractional horse power motors up to 7.5 kW than. Again, ferrites have high electric resistances, hence, no eddy current loss occurs within the PMs volume. Ferrites are produced through powder metallurgy. Some commonly available ferrites are Barium Ferrite ($BaOx6Fe_2O_3$),Strontium Ferrite, ($SrOx\ 6Fe_2O_3$) and Lead ferrite ($PbOx6Fe_2O_3$). Among them, Strontium Ferrite has higher coercive force than barium ferrites. Barium ferrites are commonly used in small dc motors used in automobiles. Lead Ferrites have some manufacturing problems regarding environmental issues.

2.2.3 Rare-Earth Magnet

The first generation rare earth permanent magnet was invented in early 1960s and was commercially produced since 1970s. Rare earth magnets are not in general rare but their composition compounds which have no other applications are need to be refined making the materials costly. The first developed rare earth material $SmCo_5$ [8] has high remanence flux density, high coercive force, high energy product, and linear demagnetisation curve and low temperature coefficient. Maximum operating temperature is 300-350°C. Due to its high cost this is suitable for motors with low volume, high power density and for H or F class of insulation. A higher value of the maximum energy product $(BH)_{max}$ of 156 kJ/m³ could be obtained through several special production technique such as pressing, liquid phase sintering etc. The second generation of SmCo, was developed as $Sm_2\,Co_{17}$ which could have a $(BH)_{max}$ as high as 238 kJ/m³.

In 1983 Sumitomo Special Metals, Japan announced a new generation of Permanent magnet NdFeB at the 29th Annual Conference of Magnetism and Magnetic Materials held at Pittsburgh, USA. NdFeB shows a better magnetic properties than SmCo at room temperature. In 2007, a NdFeB is reported with a $(BH)_{max}$ as high as 415 kJ/m³. But, the negative temperature coefficient of B_r (-0.09 to -0.15% /°C) and H_c (-0.4 to -0.8% /°C) is much higher than SmCo which limits its operation at higher temperature. Maximum operating temperature of NdFeB is 250°C. NdFeB also is susceptible to corrosion. Still as its performance-to-cost ratio could be enhanced largely, it is suitably used for many PM apparatus. The demagnetisation characteristics of different magnet material have been shown in Figure 2.4.

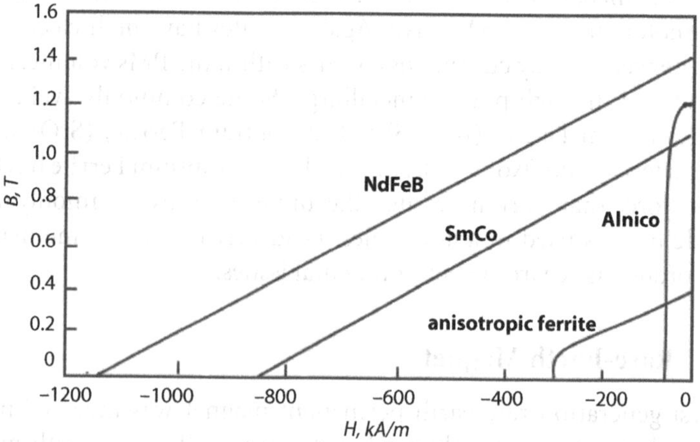

Figure 2.4 Demagnetisation curve of Alnico, Ferrite, Nd-Fe-B and SmCo at 20°C.

2.3 Direction of Magnetisation

Magnetisation direction of permanent magnet will play an important role for the distribution of magnetic flux density along the air gap periphery as well as affects the power density indirectly for any type of permanent magnet synchronous motor. Air gap flux density even has large impact on the generation of harmonic torque which in turn distorts the quality of output torque of the machine. Mainly four types of magnetisation is there viz. radial, parallel, radial sine magnetisation and sine angle or sine direction magnetisation (Figure 2.5). Among them parallel and radial magnetisation are commonly used although other two have their unique features in some particular cases.

In case of radial magnetisation, the magnetisation is along the radius where as in parallel magnetisation, direction of magnetisation is parallel to the surface edge as shown in Figure 2.6. Magnetisation vector is shown by **M** and vector **n** represents the normal direction to the surface.

For parallel magnetized PMSM the air gap flux density is sinusoidal in nature whereas the radial magnetization will produce a rectangular flux density along the air gap periphery. The flux and the flux density distribution for a 2 pole radially and parallelly magnetized inset type PMSM has been shown in Figure 2.7. For radially magnetized magnets the flux density vector **M** is normal along the radial direction which results maximum flux to flow in the radial direction. Due to this the flux density is maximum along the air gap and remains constant throughout the magnet surface. At the same time for parallel magnetized magnets the flux that enters the stator is proportional to the sine of the angle made by the magnetisation

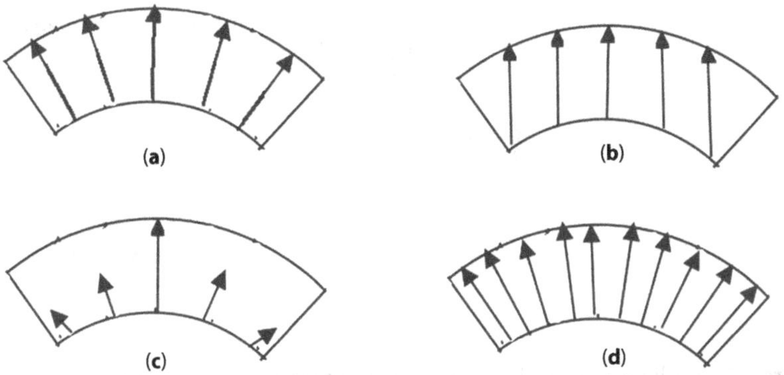

Figure 2.5 (a) radial magnetization; (b) parallel magnetization; (c) radial sine magnetization; (d) Sine angle or sine direction magnetization.

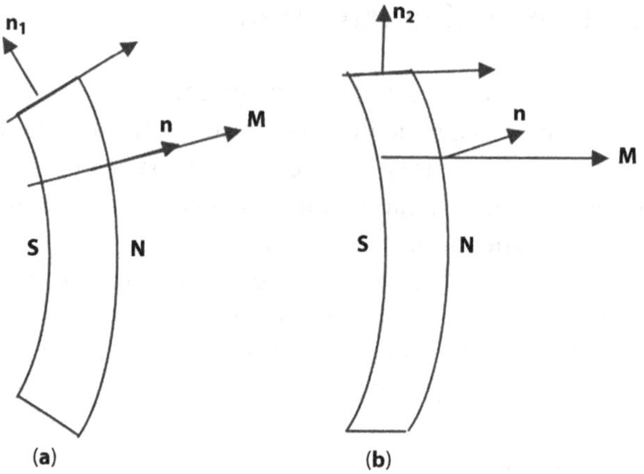

Figure 2.6 direction of magnetization of magnets (a) radial magnetization and (b) parallel magnetization.

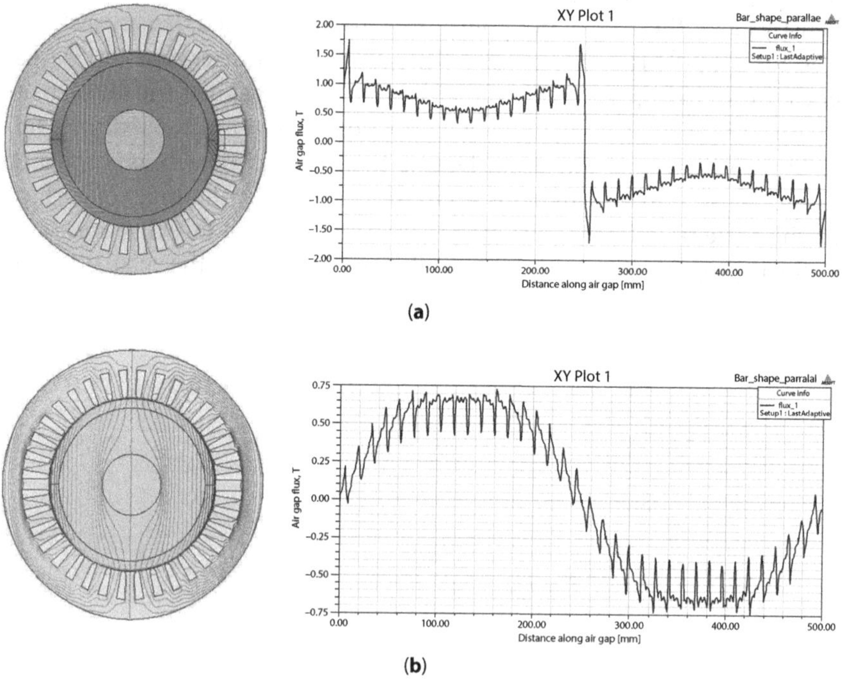

Figure 2.7 Flux lines and flux density with respect to distance along air gap periphery (a) for radially magnetised PM machine and (b) for parallelly magnetised PM machine.

vector **M** with the x-axis. This results a sinusoidal distribution of the air gap flux density.

There is another arrangement of magnets where radial or parallel direction of magnetisation is not the prime factor rather the magnets are arranged in a particular array to obtain the magnetic flux along a prescribed direction. This arrangement is called halbach array or halbach arrangement [9]–[18]. Figure 2.8 shows the halbach array by using two magnet array, one is radial magnet array and other is azimuthal magnet array. Combination of this two will give a halbach array which have a flux either in clockwise or in anti-clockwise direction as shown in Figure 2.9.

A FEA analysis of halbach array has been shown in Figure 2.10 which shows the fluxes in a particular direction. The figure shows that leakage fluxes on the upper side of the magnet arrangement is minimum hence on upper side the back iron is not essential and any nonferrous material like aluminium or other with good structural features can be used. As the fluxes are concentrated in one side, hence for inner rotor or outer rotor PMSM different arrangement has to be used.

2.4 Permanent Magnet Synchronous Motors

Permanent Magnet Synchronous Motors are basically a synchronous motor with permanent magnet rotor. Absence of the winding in the rotor circuit reduces the rotor copper losses, rotor inertia and at the same time increases the power density, makes the system more robust. The only drawback is the

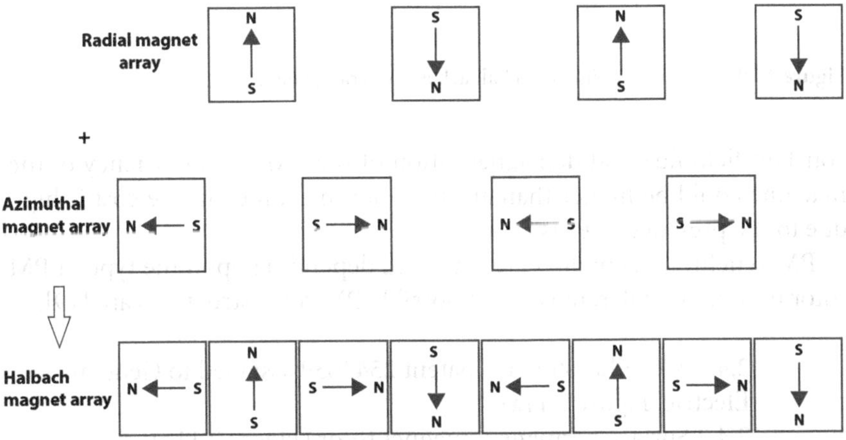

Figure 2.8 Halbach array and halbach arrangement.

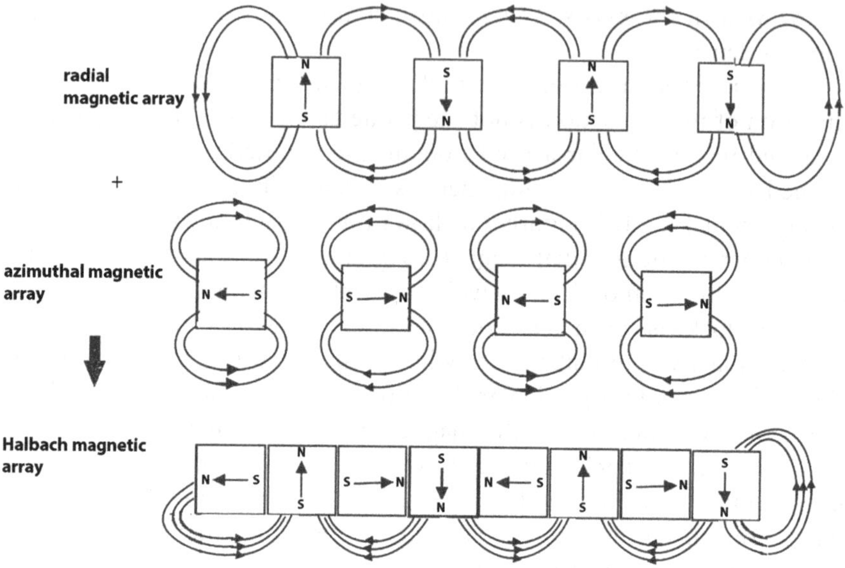

Figure 2.9 Distribution of flux in halbach array.

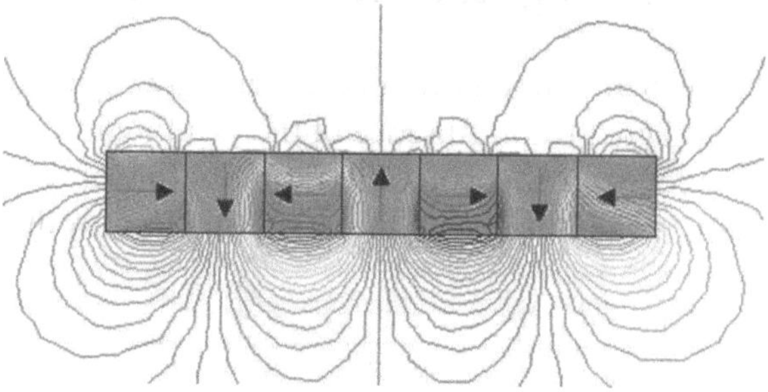

Figure 2.10 Directions of flux for a halbach array arrangement.

constant field flux and demagnetisation of the PMs. The efficiency of the machine could be higher than an Induction machine but the cost is high due to the presence of PMs.

PM synchronous motors are classified depending upon the type of PM rotor used in it. Different types of possible PM rotor structures are [19]:

2.4.1 Merrill's rotor (US patent 2543639 assigned to General Electric, Figure 2.11a)

2.4.2 Surface permanent magnet rotor (Figure 2.11b)

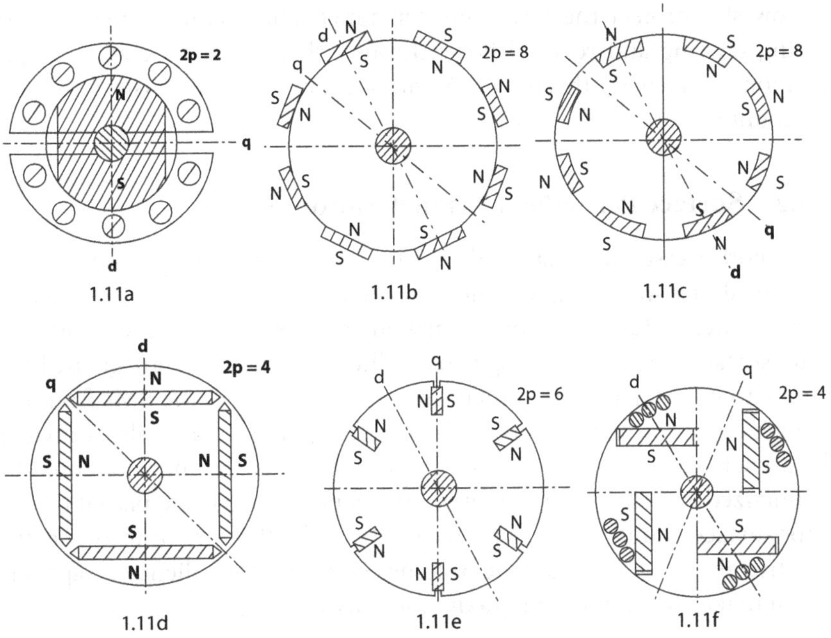

Figure 2.11 (a) Merrill's Rotor (b) Surface Permanent Magnet Motor (c) Inset Permanent magnet Rotor (d) Interior Permanent Magnet Rotor (e) Rotor with symmetrically distributed magnets (f) Rotors with magnets distributed asymmetrically.

2.4.3 Inset permanent magnet rotor (Figure 2.11.c)
2.4.4 Interior permanent magnet rotor (Figure 2.11.d)
2.4.5 Rotor with symmetrically distributed magnets (Figure 2.11.e)
2.4.6 Rotors with magnets distributed asymmetrically (also known as *Siemosyn* as per the German patent 1173178 assigned to Siemens, (Figure 2.11.f)

2.4.1 Merrill's Rotor

F.W. Merrill was the first to develop a PM excited synchronous motor in 1955, with alnico and ferrite alloys for small synchronous motor at high frequencies (US patent 2543639) [20]. It was a four pole motor, with the cage winding in the rotor which made it self-starting. The width of deep narrow slots within each pole on the laminated external ring can adjust the leakage flux produced by the PM. This laminated ring along with the narrow slot protect the PM from demagnetisation as the armature flux during starting and reversal flows through them. The reluctance torque component is low for this motor. An aluminium or zinc alloy sleeve is used to mount the PM on the shaft.

2.4.2 Surface Permanent Magnet Motor (SPM)

SPM motors essentially have a three phase stator winding to produce synchronously rotating magnetic field in the air gap along with either solid or laminated rotor core. The permanent magnets are attached with the outer surface of the rotor using epoxy adhesive. A non-ferromagnetic high conductivity external ring sometimes is used to protect the PMs from the demagnetizing armature flux and centrifugal sources and also provides the asynchronous starting torque. The PMs for this kind of rotor could be magnetized radially or circumferentially. Since the PMs are placed on the rotor surface and the relative permeability of the PMs are near to one, the effective air gap of the machine becomes large and the saliency disappears which makes the d-axis and q-axis reactances equal.

2.4.3 Inset Permanent Magnet Motor

In inset PM motor, the PMs are placed in the inner periphery of the rotor core. The rotor core could be made of laminated sheets or solid steel. The PMs essentially radially magnetized in this type of rotor. The stator has conventional three phase winding for producing rotating magnetic field. The q-axis magnetic flux can pass through the steel pole pieces without crossing the PMs, which makes the effective air gap along the q-axis lower than that along the d-axis. This makes the machine a salient pole type with $X_{dm} < X_{qm}$. Due to the lower effective air gap the effect of armature reaction is prominent in this type of PM motors and the induced emf is also lower than that in case of SPM motors.

2.4.4 Interior Permanent Magnet Motor

Interior Permanent Magnet motor has radially magnetized PMs placed inside the narrow sleeves within the rotor. As the pole area on the rotor surface is higher than the magnet pole area, the open circuit air gap flux density is less than the magnet flux density [21]. The d-axis synchronous reactance is lower than the q-axis synchronous reactance ($X_d < X_q$) like the inset type. Due to its structure, this kind of PM motors are more robust than a SPM, allowing a much high frequency and high speed operation.

2.4.5 Buried Permanent Magnet Motor

In a buried permanent magnet motor the permanent magnets are circumferentially magnetized and embedded in to the deep slots within the rotor core. The height of the permanent magnet is along the pole pitch due to the circumferential magnetisation. The q-axis synchronous reactance is higher than the d-axis synchronous reactance ($X_q > X_d$). The width of the magnetic material between the inner sides of two consecutive permanent magnets should be selected carefully. The non-ferromagnetic shaft is essential as in case of ferromagnetic shaft a large portion of the magnetic flux will flow through the shaft reducing the useful flux at a lower value. A non-ferromagnetic sleeve between the shaft and rotor core could be used as flux barrier.

2.5 Comparison of PMSM with Induction Motor

PMSMs are more efficient than induction motor due to the absence of rotor winding losses. A comparison between PMSM and Induction motor has been given in Table 2.1 with respect to speed, starting torque, power factor, air gap and cost.

Table 2.1 Comparison between synchronous motor and induction motor.

Quantity	Synchronous motor	Induction motor
Speed	Constant speed, independent of load	Speed decreases as the load increases
Starting torque	Does not have any inherent starting torque	Inherent starting torque
Air gap	From a fraction of mm to a few mm	As small as possible
Power factor	Can be adjusted with drive, UPF operation could be possible	Dependable on the air gap, pf varies from 0.1 at no load to 0.8, 0.9 at rated load
Torque-voltage characteristics	Torque is directly proportional to the voltage	Torque is directly proportional to the square of the voltage
Cost	More expensive than Induction motor	Cost effective

2.6 Operating Principle of PMSM

PMSM has a stator with three phase winding displaced by 120° electrical in space and a rotor with PM. When a three phase AC voltage is applied across the three terminals of the stator winding a synchronously rotating magnetic field is produced. Permanent Magnet in the rotor will provide the required field flux. An EMF will be induced in the stator winding due to the presence of this magnetic flux.

2.6.1 Induced EMF

Rms value of the per phase no load induced emf in the stator winding is given by,

$$E_f = \sqrt{2}\pi f T_{ph1} k_{\omega 1} \varphi_{f1} \tag{2.1}$$

where T_{ph1} is the per phase no of turns of stator, k_{w1} is the stator winding factor which is the product of the pitch factor k_p and the distribution factor k_d and ϕ_{f1} is the fundamental component of the permanent magnet flux ϕ_f without armature reaction which is given by

$$\varphi_{f1} = L_i \int_0^{\tau} B_{mg1} sin\left(\frac{\pi}{\tau} x\right) dx = \frac{2}{\pi} \tau L_i B_{mg1} \tag{2.2}$$

where, L_i is the internal length of the core, B_{mg1} is the maximum value of the fundamental component of the air gap flux density B_{mg} and τ is the pole pitch.

Similarly, due to the d-axis and q-axis armature reaction flux ϕ_{ad} and $\phi_{aq,}$ emf E_{ad} and E_{aq} will be induced across the armature winding, which could be obtained as

$$E_{ad} = \sqrt{2}\pi f T_{ph1} k_{\omega 1} \varphi_{ad} \tag{2.3}$$

$$E_{aq} = \sqrt{2}\pi f T_{ph1} k_{\omega 1} \varphi_{aq} \tag{2.4}$$

Where, ϕ_{ad} and ϕ_{aq} are $\varphi_{ad} = \frac{2}{\pi} \tau L_i B_{mad1}$ and $\varphi_{aq} = \frac{2}{\pi} \tau L_i B_{maq1}$. Here, B_{mad1} and B_{maq1} are the peak values of the fundamental component of the d- axis and q-axis armature flux density.

2.6.2 Electromagnetic Power

The electromagnetic power developed with a negligible armature resistance (r_a =0) is expressed as

$$P_{el} = m\left[\frac{V_1 E_f}{X_{sd}} + \frac{V_1^2}{2}\left(\frac{1}{X_{sq}} - \frac{1}{X_{sd}}\right) sin2\delta\right] \tag{2.5}$$

where, m is the no of phases, V_1 is the supply phase voltage, E_f is the induced emf, X_{sd} and X_{sq} are the d- and q- axis synchronous reactances respectively and δ is the load angle between E_f and V_1.

d- and q- axis synchronous reactances, X_{sd} and X_{sq} can be obtained as

$$X_{sd} = X_1 + X_{ad} \tag{2.6}$$

$$X_{sq} = X_1 + X_{aq} \tag{2.7}$$

where $X_1 = 2\pi f L_1$ is the leakage reactances and X_{ad}, X_{aq} are the d- and q-axis armature reaction reactances which are also called d- and q-axis mutual reactances respectively. The leakage reactance X_1 constitutes with slot leakage reactance, tooth-top and end-connection differential leakage reactances. Among them only slot and differential leakage reactances depend on the magnetic saturation. The d-axis mutual reactance X_{ad} depends on the magnetic circuit saturation whereas X_{aq} is dependable on the rotor configuration.

2.6.3 Electromagnetic Torque Developed

The electromagnetic torque, T_{em} in a PMSM can be obtained by dividing the developed power P_{el} with the synchronous speed ω_s and is represented as the following

$$T_{em} = \frac{P_{el}}{\omega_s} = \frac{m}{2\pi n_s}\left[\frac{V_1 E_f}{X_{sd}} sin\delta + \frac{V_1^2}{2}\left(\frac{1}{X_{sq}} - \frac{1}{X_{sd}}\right) sin2\delta\right] = T_{esyn} + T_{erul} \tag{2.8}$$

The above equation shows that the electromagnetic torque of a salient pole synchronous motor has two components. The first component T_{esyn} is known as the synchronous torque and the second component T_{erel} is known as reluctance torque and are defined as follows

$$T_{esyn} = \frac{m}{2\pi n_s}\frac{V_1 E_f}{X_{sd}} sin\delta \tag{2.9}$$

$$T_{erel} = \frac{m}{2\pi n_s}\frac{V_1^2}{2}\left(\frac{1}{X_{sq}} - \frac{1}{X_{sd}}\right) sin2\delta \tag{2.10}$$

The first component T_{esyn} is dependent on both the supply voltage, V_1 and the induced voltage, E_f whereas the second component is dependent only on the supply voltage, V_1 and exists also in an unexcited salient pole machine where $X_{sd} \neq X_{sq}$. For non-salient pole or cylindrical rotor PMSM, where $X_{sd} = X_{sq} = X_s$, the second component of the torque equation will be absent and it will be modified as

$$T_e = \frac{m}{2\pi n_s}\frac{V_1 E_f}{X_s} sin\,\delta \tag{2.11}$$

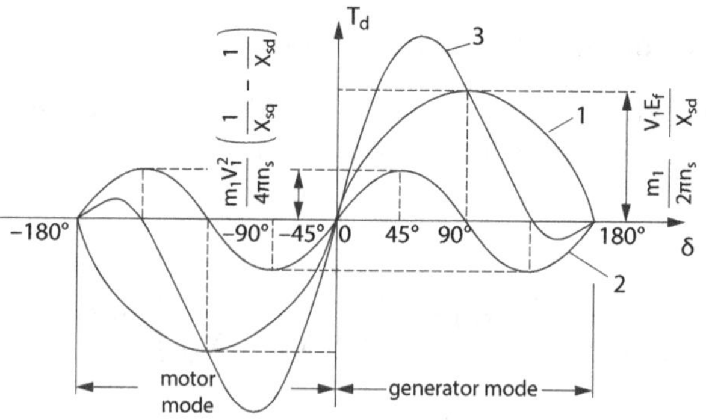

Figure 2.12 Torque-angle characteristics of a salient pole synchronous machine.

Figure 2.12 shows the variation of the toque components with the torque angle. For wound field salient pole synchronous machine $X_{sd}>X_{sq}$ where as for some special type of PMSM like 1.11.c, 1.11.d $X_{sd}<X_{sq}$.

2.6.4 Form Factors

The form factor is defined as the ratio between the fundamental component of the air gap flux density, B_{mg1} to the air gap flux density B_{mg} i.e. $k_f = \frac{B_{mg1}}{B_{mg}} = \frac{4}{\pi} sin\frac{\pi\alpha_i}{2}$ where, α_i is defined as the pole arc to pole pitch ratio and is considered as less than one.

Similarly d-axis and q-axis armature reaction form factors are also defined as the ratios between amplitude of the first harmonic component and the maximum of the normal components of the armature reaction magnetic flux densities along d- and q-axis. The d-axis armature reaction form factor, $k_{fd} = \frac{B_{ad1}}{B_{ad}}$ and q-axis armature reaction form factor, $k_{fq} = \frac{B_{aq1}}{B_{aq}}$. With electromagnetic excitation and negligible air gap k_{fd} and k_{fq} can be calculated as

$$k_{fd} = \frac{\pi\alpha_i + sin\alpha_i\pi}{\pi} \tag{2.12}$$

And,

$$k_{fq} = \frac{\pi\alpha_i - sin\alpha_i\pi}{\pi} \tag{2.13}$$

The ratio of k_{fd} to k_f and k_{fq} to k_f are defined as d-axis and q-axis reaction factors respectively and are defined by k_{ad} and k_{aq} respectively as

$$k_{ad} = \frac{k_{fd}}{k_f} \quad \text{and} \quad k_{aq} = \frac{k_{fd}}{k_f} \tag{2.14}$$

Table 2.2 gives the magnitudes of form factors k_f, k_{fd}, k_{fq} and the reaction factors k_{ad} and k_{aq} with respect to α_i for a salient pole synchronous machines.

Table 2.2 Form factors and reaction factors for PMSM.

	α_i						
Factor	**0.4**	**0.5**	**0.6**	**0.636**	**0.7**	**0.8**	**1.0**
k_f	0.748	0.900	1.030	1.071	1.134	1.211	1.273
k_{fd}	0.703	0.818	0.913	0.943	0.958	0.987	1.00
k_{fq}	0.097	0.182	0.287	0.391	0.442	0.613	1.00
k_{ad}	0.939	0.909	0.886	0.880	0.845	0.815	0.785
k_{aq}	0.129	0.202	0.279	0.365	0.389	0.505	0.785

2.6.5 Armature Reaction Reactance

The armature reaction reactances along d- and q-axis are dependent on the effect of magnetic saturation. Saturation reduces the armature reactances. The inductive reactance of the armature of a non salient pole synchronous machine is expressed as

$$X_a = 4m\mu_0 f \frac{(T_{ph}k_{w1})^2 \tau L_i}{\pi p g'} \tag{2.15}$$

Hence, the d-axis armature reactance X_{ad} and q-axis armature reactance X_{aq} considering the magnetic saturation can be calculated as

$$X_{ad} = k_{fd} X_a = 4m\mu_0 f \frac{(T_{ph}k_{w1})^2 \tau L_i}{\pi p g'} k_{fd} \tag{2.16}$$

$$X_{aq} = k_{fq} X_a = 4m\mu_0 f \frac{(T_{ph}k_{w1})^2 \tau L_i}{\pi p g_q'} k_{fq} \tag{2.17}$$

where, μ_0 is the magnetic permeability of free space, L_i is the effective length of the stator core, g' and g_q' are the equivalent air gap along d-axis and q-axis which can be represented as $g' = g k_C k_{sat} + \frac{h_m}{\mu_{rrec}}$ and $g_q' = g_q k_C k_{satq}$ where

g and g_q are the air gap length along d- and q-axis, k_C is the Carter's coefficient and $k_{sat} \geq 1$ is the saturation factor of the air gap. For salient pole rotor with electromagnetic excitation, the armature reaction flux completes its path through the larger air gap in the interpolar region along q-axis. As a result, armature reaction flux depends slightly on the saturation and hence saturation factor $k_{satq} \approx 1$.

Carter's coefficient k_c can be calculated directly from the following empirical formula:

$$k_C = \frac{1}{1+5\left(g/W_s\right)}$$ where g= air gap length and W_s =the slot width

Another empirical formula which can be used to calculate the Carter's coefficient for parallel sided open slots is as follows:

$$k_C = \frac{2}{\pi}\left[tan^{-1}y - \frac{1}{\pi}log\sqrt{(1+y^2)}\right] \text{ where } y = \frac{W_s}{2g}$$

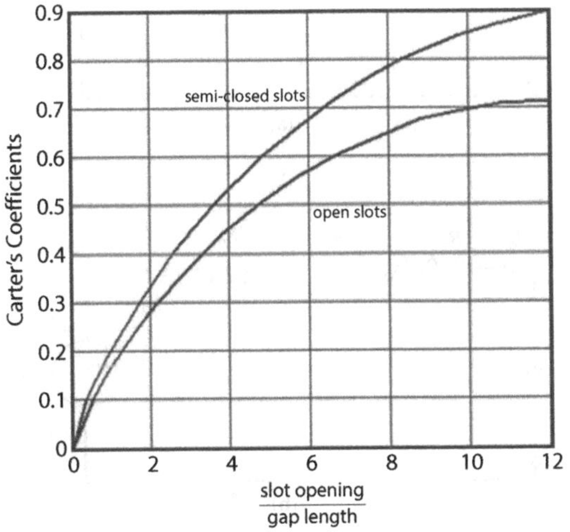

Figure 2.13 Carters coefficient for semi closed and open slots.

Carter's coefficient is dependent on the ratio of slot opening to gap length and hence is dependent on the shape of the slots. The value of Carter's coefficient for and semi closed slots and open slots can be obtained from Figure 2.13.

2.7 Equivalent Circuit

Per phase steady-state equivalent circuit of a PMSM has been shown in Figure 2.14a where R_s represents the per phase resistances, X_l is the leakage reactance, X_m is the magnetizing reactance, R_c represents the core loss resistance and E_f is the per phase induced emf across the armature terminals. Neglecting the core loss resistance, the circuit modifies as shown in Figure 2.14b where, $X_s = X_l + X_m$ represents per phase synchronous reactance and $Z_s = R_s + jX_s$ is known as the synchronous reactances. The core loss will be considered as a percentage of the output voltage with friction and windage losses as constant losses like other electrical machines.

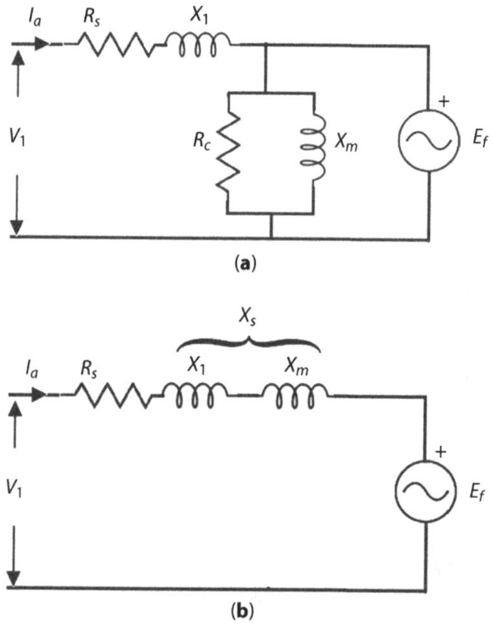

Figure 2.14 (a) Per phase equivalent circuit considering core loss branch. (b) Per phase equivalent circuit neglecting core loss branch.

2.8 Phasor Diagram

For a permanent magnet synchronous motor the voltage equation can be written as,

$$\vec{V}_1 = \vec{E}_f + \vec{I}_a R_s + j\vec{I}_d X_{sd} + j\vec{I}_q X_{sq} = \vec{E}_f + \vec{I}_a R_s + j\vec{I}_d (X_1 + X_{ad}) + j\vec{I}_q (X_1 + X_{sq}) \tag{2.18}$$

$$\text{And, } \vec{I}_a = \vec{I}_d + \vec{I}_q \tag{2.19}$$

The voltage and current phasor of a PMSM as per the Eq. 2.18 and 2.19 for underexcited and overexcited condition have been shown in Figure 2.15a and b. Synchronous motor at underexcited condition will draw required reactive power from the ac line and will operate at lagging power factor. In Figure 2.15a I_q has been shown along the E_f essentially for motoring mode. I_a lags the supply voltage V_1 by an angle θ. The internal power factor angle ψ by which I_a lags E_f is given by $\delta+\theta$ where δ is the load angle.

$$\text{Hence, } I_d = I_a \sin\psi \qquad \text{and} \quad I_q = I_a \cos\psi \tag{2.20}$$

The rms values of the d-axis and q-axis armature currents for an underexcited synchronous motor can be found out using Figure 2.16. The projections of input voltage V_1 on the d- and q-axes are

$$V_1 \sin\delta = I_q X_{sq} - I_d R_1 \tag{2.21}$$

$$V_1 \cos\delta = E_f + I_d X_{sd} + I_q R_1 \tag{2.22}$$

Whereas for an overexcited synchronous motor

$$V_1 \sin\delta = I_q X_{sq} + I_d R_1 \tag{2.23}$$

$$V_1 \cos\delta = E_f - I_d X_{sd} + I_q R_1 \tag{2.24}$$

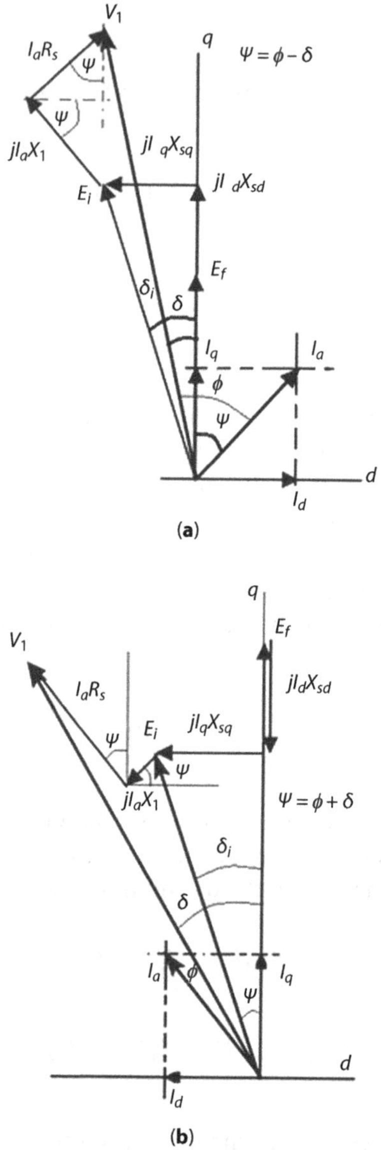

Figure 2.15 (a) Phasor diagram for underexcited synchronous motor. (b) Phasor diagram for overexcited synchronous motor.

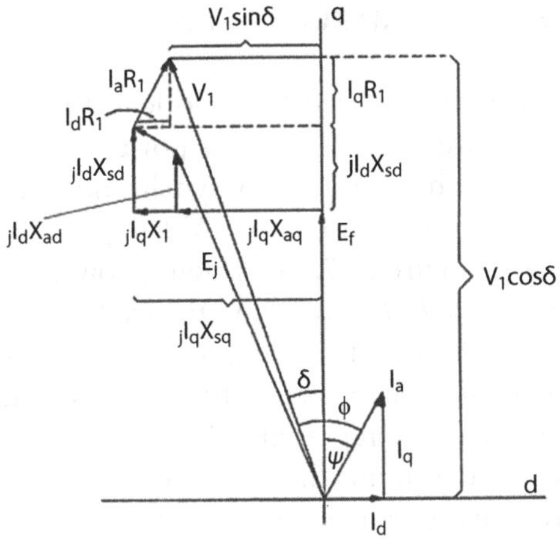

Figure 2.16 Phasor diagram to calculate I_d and I_q for underexcited motor.

By solving Eq. 2.21 and 2.22, the d-and q-axis currents for an underexcited motor have been obtained as

$$I_d = \frac{V_1(X_{sq}\cos\delta - R_1\sin\delta) - E_f X_{sq}}{X_{sd}X_{sq} + R_1^2} \tag{2.25}$$

$$I_q = \frac{V_1(R_1\cos\delta + X_{sd}\sin\delta) - E_f R_1}{X_{sd}X_{sq} + R_1^2} \tag{2.26}$$

And the same have been obtained for overexcited motor solving the Eq. 2.23 and 2.24 as,

$$I_d = \frac{V_1(R_1\sin\delta - X_{sq}\cos\delta) + E_f X_{sq}}{X_{sd}X_{sq} + R_1^2} \tag{2.27}$$

and the q-axis current Iq will be the same as Eq. 2.26.

2.9 Performance Characteristics

PMSM runs at synchronous speed from no load to full load at a power factor near to unity. Torque- load angle or power-load angle characteristics are most important characteristics of synchronous motor. The no load characteristics of synchronous motor $I_a = f(V_1)$, $P_{out} = f(V_1)$, and $cos\phi = f(V_1)$ at $T_{sh} = 0$ are shown in Figure 2.17. The lagging power factor is on the left hand side of the point $cos\phi=1$ or $\phi= 0$and the leading power factor (overexcited) is on the right hand side of the point $cos\phi=1$. That means the left hand side of the characteristic represents the underexcited condition of the field and the right hand side represents the overexcited condition of the field. The threshold voltage limits the curve for underexcited region where as maximum current in the armature winding limits the overexcited region of the curve. The overexcited synchronous motor at no load can compensate the reactive power consumed by the inductive load and is known as *synchronous condenser.*

The overload capacity is described as a ratio of maximum output power or maximum torque to rated output power or rated torque respectively. At the same time ratio of output power, P_{out} to rated output power, P_r is defined as the relative output power with respect to which the armature current I_a, Shaft torque T_{sh}, input power P_{in}, efficiency η and the power factor $cos\phi$ are plotted (Figure 2.18).

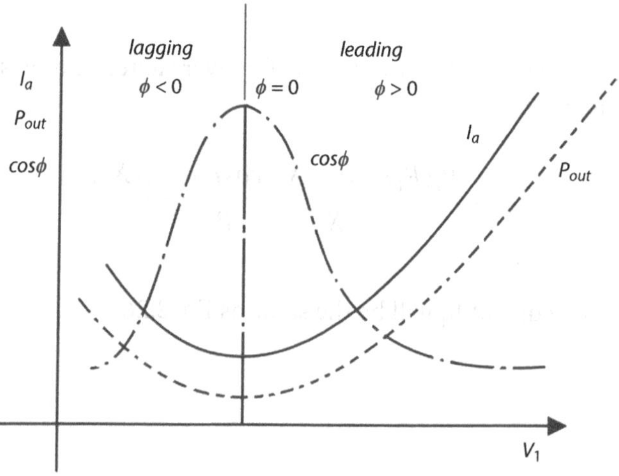

Figure 2.17 The no load characteristics: $I_a = f(V_1)$, $P_{out} = f(V_1)$, and $cos\phi = f(V_1)$.

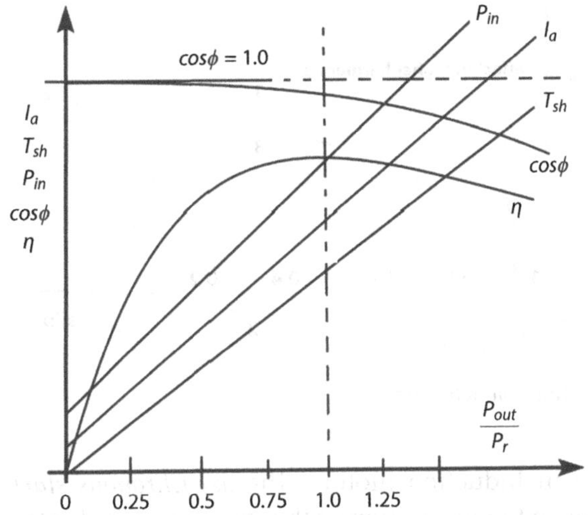

Figure 2.18 Armature current I_a, Shaft torque T_{sh}, input power P_{in}, efficiency η and the power factor $\cos\phi$ as a function of (P_{out}/P_r).

2.10 Starting

When the rotor is stationary i.e. at standstill condition the relative velocity between the stator produced rotating magnetic field and the stationary rotor conductors is equal to synchronous speed. As torque angle i.e. the angle between the rotating stator mmfs and stationary rotor mmf is a function of time, electromagnetic torque will vary sinusoidally with respect to time and reverses after each half cycle. Hence the average value of torque becomes zero over a complete cycle. As a consequence, synchronous motor does not have any inherent starting torque.

Steady electromagnetic torque can be produced if the relative velocity between the stator produced rotating magnetic field and rotor produced field becomes zero. This can happen only if rotor is being pulled at synchronous speed with some additional arrangements. Two methods are commonly adopted to start a synchronous motor. One is *auxiliary motor start* and other one is *asynchronous motor start.*

2.10.1 Asynchronous Motor Start

In *asynchronous motor starting,* damper winding is being placed in the rotor slots on the rotor pole faces. These damper winding is short circuited by both ends with conducting metal rings which is similar to the squirrel

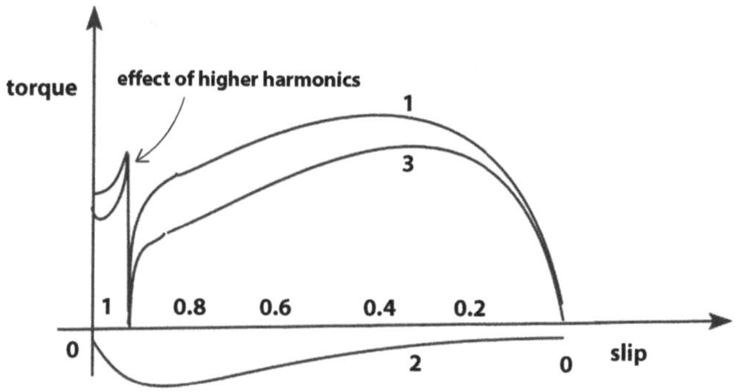

Figure 2.19 Starting characteristics.

cage winding in Induction motors. The *asynchronous starting torque* is developed due to the interaction of the stator produced rotating magnetic field and rotor currents through the damper winding. At starting permanent magnet produces a negative torque which behaves as braking torque. Proper selection of type and volume of permanent magnet is required to ensure the starting of permanent magnet synchronous motor. Figure 2.19 represents the starting characteristics of PMSM with respect to time.

PMSM which can develop inherent asynchronous starting torque, are commonly known as Line start Permanent Magnet Synchronous Motor (LSPMSM). After starting, the rotor is pulled into synchronism with the stator produced field and continues to rotate at synchronous speed. The bars in the rotor of a LSPMSM are not skewed due to the presence of PM in the rotors. Instead the stator lots can be skewed if required. LSPMSM can be designed to operate at unity power factor with super premium efficiency and it is being thought that sooner induction motor will be replaced by LSPMSM in the global market.

2.10.2 Auxiliary Motor Start

Synchronous motors can be started with the help of an auxiliary motor coupled with the shaft of the synchronous motor. Induction motors are normally used for this purpose. With the help of an auxiliary induction motor, the synchronous motor is being run at a speed near to the synchronous speed (at slip less than 2%) and first its armature winding and then the field winding is being excited. With the field winding excitation the rotor will be pulled into synchronism and will start to rotate at synchronous speed. The main drawback of this method is incapability of the synchronous motor to be started under load and use of the same rating auxiliary motor makes the installation costly.

2.10.3 Variable Frequency Start

Variable Frequency start is a common starting method of synchronous motor where the supply frequency is slowly increased from zero value to the rated value using a Variable Voltage Variable frequency (VVVF) inverter. Slow change in the input frequency will allow the rotor to be in synchronism with the stator rotating magnetic field. As the frequency is changing the speed of the rotating magnetic field is changing but the rotor is allowed to cope up with the field. Slow and smooth change in frequency is required.

2.11 Dynamic Modeling of PMSM

Dynamic performance of a PMSM could be obtained by developing the Park Model with synchronously rotating d^e- q^e reference frame rotating with a velocity ω_e. This model is necessary to develop the control strategy for PMSMs.

The dynamic equations of the Permanent Magnet Synchronous Motor can be written as:

$$v_{qs} = r_s i_{qs} + \omega_e \psi_{ds} + \frac{d\psi_{qs}}{dt} \quad (2.28)$$

$$v_{ds} = r_s i_{ds} - \omega_e \psi_{qs} + \frac{d\psi_{ds}}{dt} \quad (2.29)$$

$$0 = r_D i_D + \frac{d\psi_D}{dt} \quad (2.30)$$

$$0 = r_Q i_Q + \frac{d\psi_Q}{dt} \quad (2.31)$$

where, v_{ds}, v_{qs} are the d-and q- axis terminal voltages, r_s is the armature winding resistances, i_{ds}, i_{qs} are the d- and q-axis armature currents, i_D and i_Q

are the d-and q-axis component of damper winding currents and r_D, L_D, r_Q, L_Q are the d-axis, q-axis resistances and inductances respectively of damper winding. The d-and q-axis armature and damper winding flux linkages are given as

$$\psi_{ds} = i_{ds}(L_{dm} + L_{ls}) + L_{dm}i_D + \psi_f = i_{ds}L_{ds} + L_{dm}i_D + \psi_f \quad (2.32)$$

$$\psi_{qs} = i_{qs}(L_{qm} + L_{ls}) + L_{qm}i_Q = i_{qs}L_{qs} + L_{qm}i_Q \quad (2.33)$$

$$\psi_D = i_D(L_{dm} + L_D) + L_{dm}i_{ad} + \psi_f \quad (2.34)$$

$$\psi_Q = i_Q(L_{qm} + L_Q) + L_{qm}i_{aq} \quad (2.35)$$

The PM flux linkage ψ_f is given by, $\psi_f = L_{fd}I_f$. In case of PM excitation the field winding current I_f is given by , $I_f = H_c h_M$

Where, H_c is the coercivity and h_M is the height of the PM and, L_{fd} is the mutual inductance between the armature and field. Armature d-axis q-axis self-inductances are

$$\begin{aligned} L_{ds} &= L_{dm} + L_{ls} \\ L_{qs} &= L_{qm} + L_{ls} \end{aligned} \quad (2.36)$$

Where L_{dm} and L_{qm} are the magnetizing inductances along d-axis and q-axis respectively and L_{ls} is the per phase leakage inductances of the armature winding.

In case of no damper winding in the rotor circuit $i_D = 0$ and $i_Q = 0$ and the above voltage equations will be reduced to

$$v_{qs} = r_s i_{qs} + \omega_e \psi_{ds} + \frac{d\psi_{qs}}{dt} = (r_s + \frac{dL_{qs}}{dt})i_{qs} + \omega_e L_{ds} i_{ds} + \omega_e \psi_f \quad (2.37)$$

$$v_{ds} = r_s i_{ds} - \omega_e \psi_{qs} + \frac{d\psi_{ds}}{dt} = (r_s + \frac{dL_{ds}}{dt})i_{ds} - \omega_e L_{qs} i_{qs} \quad (2.38)$$

The matrix form of the voltage equations can be written as

$$\begin{bmatrix} v_{ds} \\ v_{qs} \end{bmatrix} = \begin{bmatrix} r_s + \dfrac{dL_{sd}}{dt} & -\omega_e L_{sq} \\ \omega_e L_{sd} & r_s + \dfrac{dL_{sq}}{dt} \end{bmatrix} \begin{bmatrix} i_{ds} \\ i_{qs} \end{bmatrix} + \begin{bmatrix} 0 \\ \omega_e \psi_f \end{bmatrix} \tag{2.39}$$

The torque equation is

$$T_e = \frac{3}{2}\left(\frac{P}{2}\right)\left(\psi_{ds} i_{qs} + \psi_{qs} i_{ds}\right) = \frac{3}{2}\left(\frac{P}{2}\right)\left[\psi_f + \left(L_{sd} - L_{sq}\right) i_{ds}\right] i_{qs} \tag{2.40}$$

The d axis and q axis currents i_{ds} and i_{qs} are related with the line currents i_a, i_b and i_c as:

$$i_{ds} = \frac{2}{3}\left(i_a \cos\omega t + i_b \cos\left(\omega t - \frac{2\pi}{3}\right) + i_c \cos\left(\omega t + \frac{2\pi}{3}\right)\right) \tag{2.41}$$

$$i_{qs} = -\frac{2}{3}\left(i_a \sin\omega t + i_b \sin\left(\omega t - \frac{2\pi}{3}\right) + i_c \sin\left(\omega t + \frac{2\pi}{3}\right)\right) \tag{2.42}$$

2.12 Cogging Torque

As permanent magnet is present in the rotor, a torque is produced due to the interaction of the stator teeth and permanent magnet field, even if the stator winding is not being excited. When the PM rotor crosses the stator tooth, the total energy stored in the air gap changes with respect to the rotor angular position. This change in the total energy causes a net torque which is known as cogging torque. Although its value is small compare to full load torque, it produces a significant torque ripple and speed rippleat lower speed. This torque is also known as detent torque or no current torque. It is alternating in nature and its frequency, f_c depends upon the number of stator slots and the speed of the rotor structure as, $f_c = S_s.n_s$ [22]

The production of cogging torque can be explained using Figure 2.20.

An electrical machine having two teeth in stator and two permanent magnets in rotor has been considered. When rotor is at reference position *a*, rotor is said to be at detent position. This is not a stable position. At this position reluctance experienced by the rotor structure is maximum as the air gap is maximum. Hence, it will try to align with the position where reluctance is minimum i.e. along the stator teeth at point *b* (90° from the reference). Tendency of the rotor to be aligned with the minimum reluctance position will produce the cogging torque. This torque will increase from the reference position and will become maximum near to 45° and then will start to reduce. At *b* the torque will be zero. It is also called the detent position but is a stable one. This one is a stable position because rotor can stay here for infinite time until unless external force has been applied. To move the rotor from this position external force required. Hence the cogging torque for next 90° can be considered as negative and at point *c* again the cogging torque will be zero. The nature of the cogging torque has been shown in Figure 2.21.

Figure 2.21 reveals that the number of cycles of cogging torque for one revolution of the rotor is 2 for the above mentioned machine. With the higher frequency of the cogging torque, magnitude of the torque will reduce hence a higher order of the cogging cycles required. The no of cogging cycles depends on no of stator slots, no of rotor poles and type of the winding. It has been seen that machine with concentric winding in stator will have higher no of the cogging cycles.

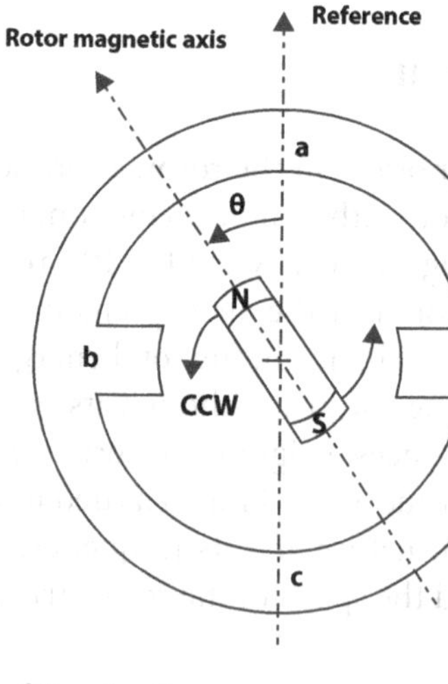

Figure 2.20 Production of Cogging Torque.

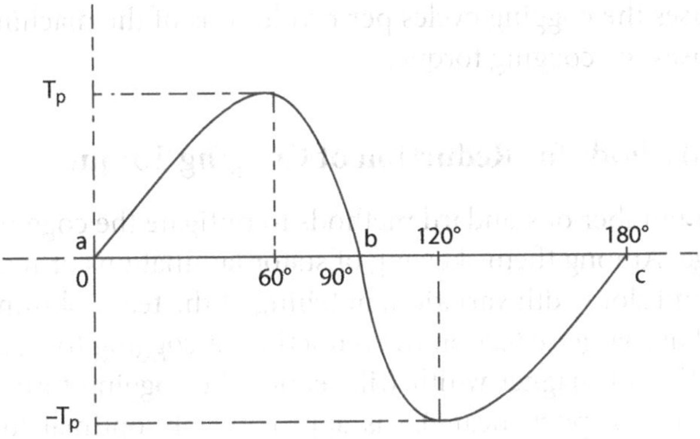

Figure 2.21 Cogging torque with respect to rotor position.

2.12.1 Reasons Affecting Cogging Torque

Cogging torques are mostly affected by the design parameters like width of stator slot, magnet strengthened manufacturing issues like manufacturing irregularities, stator laminations punching, stampings etc. Following section will discuss briefly their impacts on cogging torque:

Stator slot width: Size of stator slots not only have impact on the output of the machine, cogging torque is also produced by the interaction between the magnetic field and the stator slots. Dimension of stator slot mainly slot opening plays an important role in generation of cogging torque. For closed slot the cogging torque is zero but it is not a common practice for all type of electrical machine to use closed slot. A minimum slot opening is required for the placement of windings in the slots. Hence, an optimized opening is to be determined for minimum cogging torque. The slot opening can be covered by magnetic pieces after placing the windings into the slots.
Magnet Strength: Remanent flux density of magnet has direct impact on the cogging torque. Temperature and initial magnetisation are the two prime factors which affects the remanent flux density directly and hence affects the cogging torque. Reduction of remanent flux density not only reduces the cogging torque but also reduces the electromagnetic torque of the machine.
Manufacturing Irregularities: Accurate positioning of permanent magnet in the rotors affects the cogging torque more prominently than the dimension of the permanent magnet. Orientation of grain, stamping, punching, welding during stacking etc introduces anisotropies in the stator laminations.

This increases the cogging cycles per revolutions of the machine which in turn increases the cogging torque.

2.12.2 Methods for Reduction of Cogging Torque

There are a number of standard methods to mitigate the cogging torque at design stage. Among them skewing of stator laminations or magnet, magnet width and slot width variation, notching of the teeth, shifting alternate pole pairs have larger effect on the reduction of cogging torque. Variation of slot width and magnet width will reduce the cogging torque. Ratio of slot width to slot pitch near 0.5 is approximately optimal to maximize the output torque. At the same time the magnet width will be selected to reduce the cogging torque.

Skewing: Skewing of stator laminations by one slot pitch in axial direction eliminates the cogging torque [23]. The skew angle to eliminate the cogging torque is given as

$$\theta_{sk} = \frac{360}{N_{c1}} \text{degree electrical} \tag{2.43}$$

where, N_{cl} is the no of cycles of cogging torque. Instead of skewing the stator laminations, PM in the rotors can also be skewed. This method makes the manufacturing process more critical hence normally it is avoided. For skewing the PMs, the pole should have multiple magnets along the axial length of the machine.

Notching of Teeth: Stator teeth are sometimes notched to create dummy teeth of minimum depth (Figure 2.22). This will increase the interaction between the permanent magnet field and stator dummy teeth and produces their own cogging torque cycles. These cogging torque cycles basically superimposed on the cogging torque cycles of the main conductor carrying slots and produces resultant cogging torque cycles. The frequency of the resultant cogging torque cycles will increase hence the magnitude of cogging torque will reduce. The selection of the notching slots should be proper otherwise the cogging torque may increase instead of decreasing. It has been seen that to mitigate the cogging torque using dummy slots, the highest common factor between the number of cogging cycles during the rotation of one slot pitch, N_p and (N_n+1) where N_n is the number of notches per stator tooth must be equal to one.

Instead of having notch on stator teeth, dummy slots can be created in between the regular slots. Alternate slot can be made free of winding to

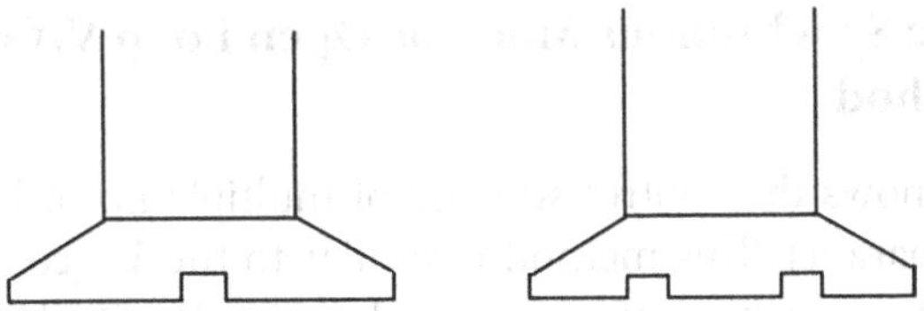

Figure 2.22 Notching of stator teeth.

produce a dummy teeth. But this will reduce the slot volume for winding. The dummy teeth may not be of same size of the regular teeth. Introduction of this kind of dummy teeth also increase the direct axis reactances and is a common practice in manufacturing of electrical machines.

2.13 Control of PMSM

Synchronous motors with permanent magnet rotors are generally used in low to medium power applications where power demand could be up to several horsepower. The drives required for synchronous motors are costlier than that used for induction motors but at the same time synchronous motor drives can run at a higher power factor level with an increased efficiency which reduces its running cost. Line start permanent magnet synchronous motors associated with damper winding can develop inherent asynchronous starting torque whereas PMSM without damper winding requires a drive for starting also. PMSM drives are commonly used in electric vehicles, ship propulsion, robotics and servo applications, cement mills, rolling mills, MAGLEV, starters or generators for aircraft engines.

As synchronous speed is related with frequency, the speed of PMSM is dependent on the inverter or cycloconverter frequency supplied by the PMSM drive. Two commonly used control methods are: true synchronous machine mode or open loop control method and self-control mode. In true synchronous machine mode, converter frequency is independently controlled to control the speed of the motor. Motor speed is controlled by the drive irrespective of actual speed of the motor. That's why this method is known as open loop control method.

Whereas in self-control method converter frequency control pulse is generated by the rotor position sensor mount on the motor shaft. This method takes care of actual speed of the machine and hence this one is a closed loop control method.

2.13.1 True Synchronous Mode or Open Loop V/f Control Method

Figure 2.23a shows the control strategy of multiple no of PMSM motors connected in parallel. This method is similar to the *V/f* control of induction motor drive. It is the simplest method of speed control but at the same time its performance is not up to the mark as vector control. This method is also known as scalar control method and is used for those applications where close tracking of speed is required among a large no of machines like fiber spinning mills, paper mills and textile mills.

The 3-phase PWM inverter with which all the synchronous motors are connected in parallel gets two control signals. One is the reference speed, ω_e^* corresponding to what all the motors will run in synchronism and the other is the reference voltage, V_s^* which is generated using a separate oscillator essentially in proportion to the frequency to keep the stator flux ψ_s constant. Constant stator flux will assure a steady electromagnetic torque and fast transient response. The frequency will be changed slowly so that rotor can track the change in speed. At the required speed the rotor will pull into step through hunting due to the presence of damper winding in the rotor. This allows the method to be suitable for starting as well as for regenerative braking also. The inverter is fed from 1 phase or 3 phase AC supply through a diode bridge rectifier and a LC filter.

The drive performance in motoring as well as in braking mode has been explained in Figure 2.23b. Initially motor has been started under no load condition from standstill position ***O*** and then frequency is gradually increasing till point ***M***. Now, consider the load torque, T_L is slowly increased. Motor steady state operating point will shift vertically to point ***N*** where motor load torque is equal with the electromagnetic torque i.e. T_L=*Te*. Developed torque, T_e is dependent on the load angle δ and stator current I_s. Hence, with constant ψ_s, both δ and I_s will increase till the next steady state point ***N*** reaches. At ***N***, the limit of either load angle (π/2) or stator current (rated current) will reach, whichever is earlier. Normally, current limit of inverter is reached earlier than machine stability limit. Keeping T_L constant if the frequency is increased slowly, the operating point will be shifted from ***N*** to point ***P***. The reduction in the load torque will bring the operating point again down to point ***Q***. For negative torque motor will enter into the braking mode and the machine speed torque profile will be along ***Q-R-S-A***. During the braking operation the recovered kinetic energy will be exhausted in the dynamic brake block **DB** installed in the dc link.

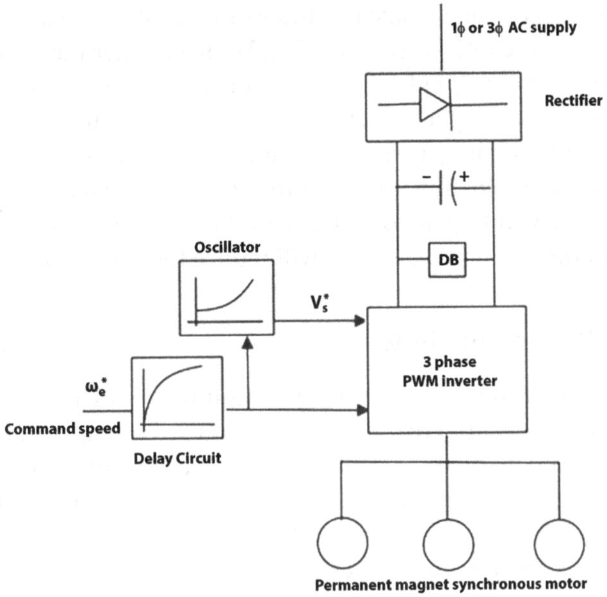

(a)

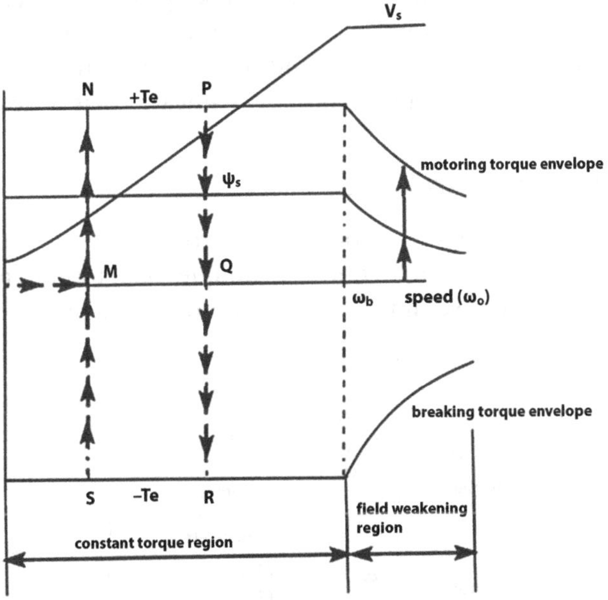

(b)

Figure 2.23 (a) Speed control of multiple PMSM through true synchronous mode. (b) V/f speed control characteristics in speed torque plane.

At the base speed the voltage reaches its rated value. Beyond it, voltage increment above the rated is not possible. With a further increment in frequency the machine will enter into the field control region. The field flux will be reduced resulting a reduction in the maximum torque. To reverse the direction of rotation, inverter terminals phase sequences has to be interchanged. Although any kind of rotor oscillations have been compensated by damper winding, it contributes to the inverter harmonic content. Increment in the harmonic currents will reduce the drive efficiency.

2.13.2 Self-Control Mode

Self-control synchronous motor drive also requires VVVF inverter like true synchronous mode. But here the frequency and the voltage commands are not controlled independently rather they will be controlled by a position sensor placed on the machine shaft showed in Figure 2.24. In self-control mode the PMSM behaves like a dc motor. Commutator and brushes in dc machines are replaced here by an electronic commutator. Hence the losses associated with the mechanical commutator as well as several disadvantages regarding maintenance, reliability, sparking, difficulty in operation in hazardous area like mines and explosive environment, limitations in power rating etc can be avoided. Hunting and stability problem also reduces due to self-control. The transient response is identical to that of dc machine.

2.13.3 Vector Control

The above mentioned methods of speed controls are scalar control method as mentioned earlier. In these control techniques both torque and flux control give a slow response and the system becomes susceptible to instability

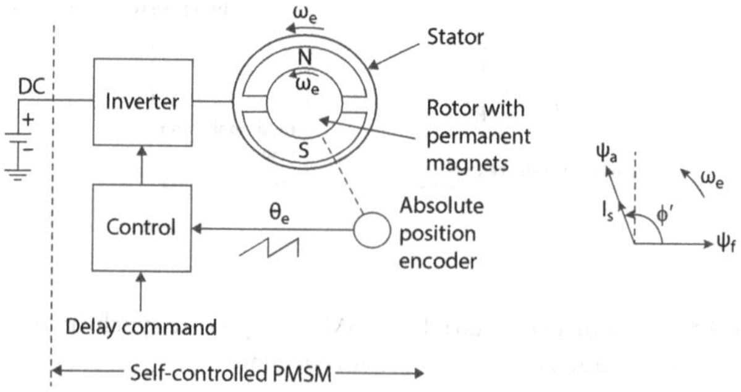

Figure 2.24 Self-controlled PMSM drive with rotor position sensor.

very quickly due to higher order system effect. This problem can be dissolved easily in vector or field oriented control where the ac machine can be controlled like a dc separately excited machine.

Figure 2.25 explains the vector control techniques in general briefly. Here, the total model has been divided into two basic parts: left side represents the control technique model and the right side describes the machine model. In machine part first machine 3ϕ terminal currents i_a, i_b and i_c are transformed in to 2ϕ currents i_{ds}^s and i_{qs}^s through linear transformation in stationary reference frame. Then in next block they are transformed into synchronously rotating reference frame using $cos\theta_e$ and $sin\theta_e$. This transformation should assure the alignment of i_{qs} along ψ_r *and* i_{ds} in quadrature with i_{qs}. In controller part the i_{ds}^* and i_{qs}^*corresponds to machine current i_{ds} *and* i_{qs}. An inverse transformation is conducted to obtain the reference machine currents i_a^*, $i_b^*$and i_c^*.These currents will be processed through an inverter with unit gain. As this transformation and linear transformation ideally do not have any dynamics the response are instantaneous.

In following topic, vector control for a sinusoidal Surface mount Permanent Magnet Synchronous Motor (SPMSM) has been discussed. In SPM as the effective length of air gap is large, magnitude of synchronous reactance (X_s) as well as armature flux linkage ψ_a is low.

$$\text{Therefore, } \vec{\psi}_s = \vec{\psi}_a + \vec{\psi}_f \approx \vec{\psi}_f \tag{2.44}$$

where ψ_s = stator flux = stator flux linkage and ψ_f = field flux linkage as shown in Figure 2.26.

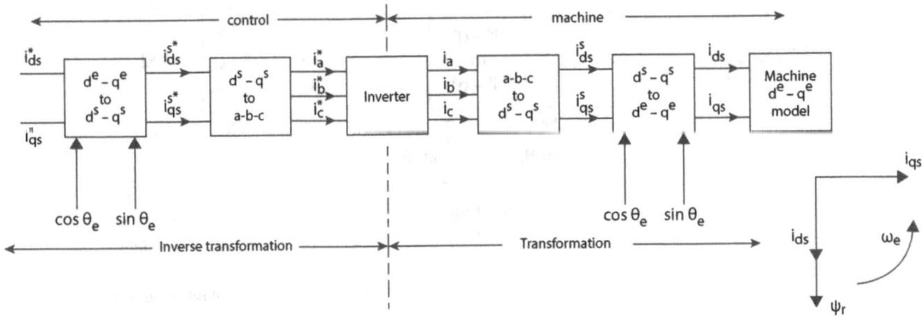

Figure 2.25 Principle of vector control.

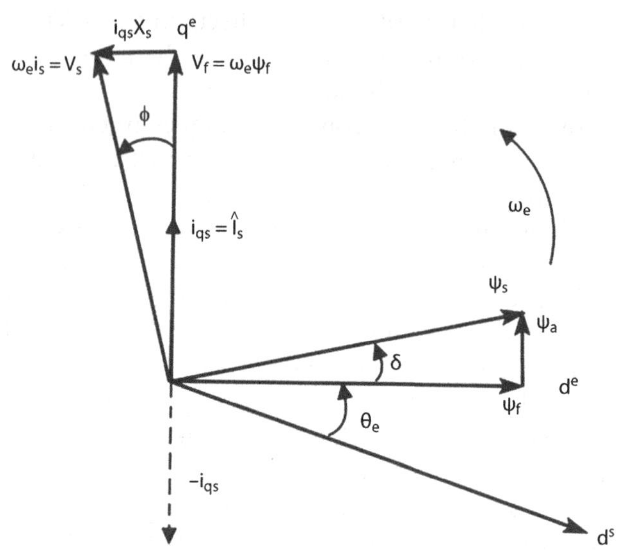

Figure 2.26 Vector control phasor diagram.

Hence, neglecting stator winding resistance condition for maximum efficiency is, i_{ds}=0 which will give

$$\vec{I}_s = i_{ds} + i_{qs} = i_{qs} \,. \tag{2.45}$$

Vector control for SPM is shown in Figure 2.27. The speed control loop generates i_{qs}^*. It will have a positive polarity for motoring mode where as a negative polarity for generating mode. The magnetizing current i_{ds}=0 as the

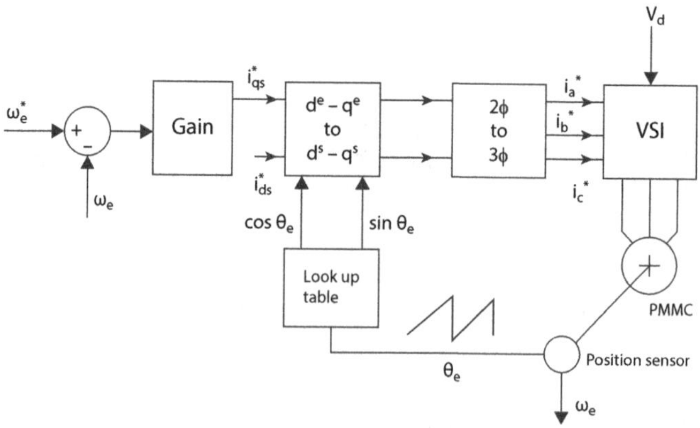

Figure 2.27 Vector control of sinusoidal SPM machine.

machine has a permanent magnet to supply the rotor flux. A position sensor is used to track the rotor movement whose output will give $cos\theta_e$ and $sin\theta_e$. Inverter reference current commands i_a^*, $i_b^*$and $i_c^*$are developed using unit vector signals ($cos\theta_e$, $sin\theta_e$) and the current signal i_{qs}^* This method is valid only for constant torque region and at the end of the constant torque region the controller will saturate as V_s *and* V_f increases with speed ω_e.

2.13.4 Field Weakening Mode

As far the discussion, we have seen that the speed bellow the rated value is controlled by armature voltage control method and at the same time, the speed above the rated is controlled by weakening the field flux. During the field flux control power remains constant and hence is called as constant power operation. However, the range of field flux control is limited by the weak armature reaction effect which lead to a minor region for field controlling.

Beyond the base speed, ω_b stator flux ψ_s must be reduced to have control over stator current to obtain higher speed. This needs $-i_{ds}$ to be injected on stator side. But due to the lower value of armature reaction flux, ψ_a the reduction in stator flux ψ_s will be low which allows a narrow zone of flux weakening. In phasor diagram (Figure 2.28) it has been depicted.

Now with $-i_{ds}$, the stator current, $\vec{I}_s = \sqrt{i_{ds}^2 + i_{qs}^2}$ and its locus varies along the line **m-m**′. For a negative i_{ds} the armature reaction flux is demagnetizing

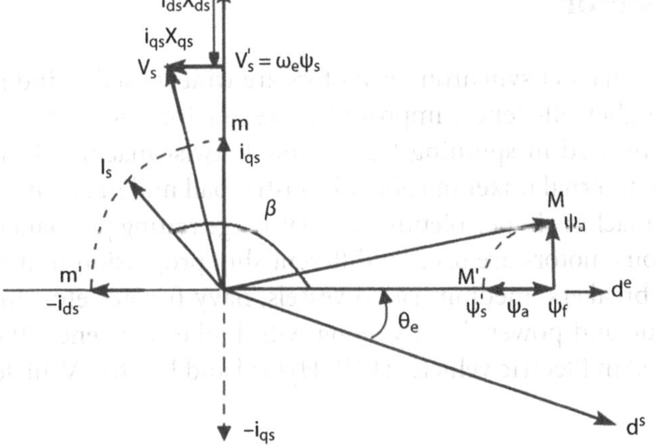

Figure 2.28 Phasor diagram for field weakening operation.

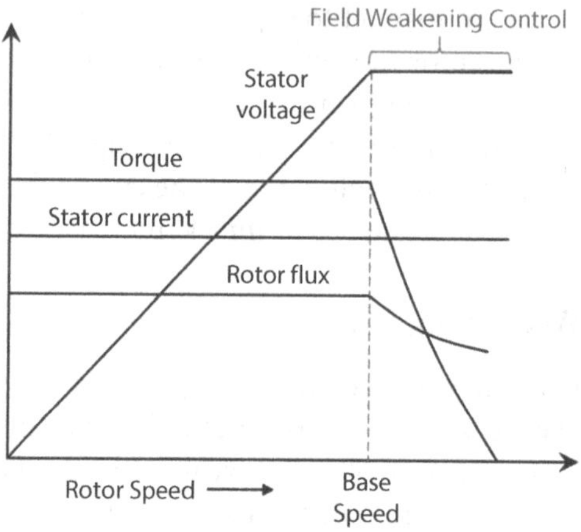

Figure 2.29 Torque-speed curve for field weakening operation.

in nature. For a constant value of ψ_a, the stator flux ψ_s will be reduced along **M-M′** and the speed will increase. For a constant field flux ψ_f, V_f should be increased beyond its rated value in proportion to the speed ω_e resulting the machine to be operated into the leading power factor region. When $\vec{I}_s = -i_{ds}$ at point m′, developed torque will be zero and speed will be maximum. It is obvious that the maximum possible speed can be increased if synchronous inductance, L_s can be increased up to L_s' within the rated stator current (Figure 2.29).

2.14 Application of Permanent Magnet Synchronous Motor

Permanent magnet synchronous motors are widely used in industries due to their higher efficiency, improved power factor and simple structure. PMSMs are used in spinning frame, double twist machine in the textile industries, internal mixer in rubber industry, ball mill in ceramic industry, pumping machine in petroleum industry. Large ratting permanent magnet synchronous motors are used in different ship propulsion unit like cruise ships, ice breakers, medium cargo vessels, navy frigates etc. Due to their high torque and power density along with higher efficiency, PMSMs are mostly used in Electric vehicles (EV), Hybrid and Electric Vehicles (HEV).

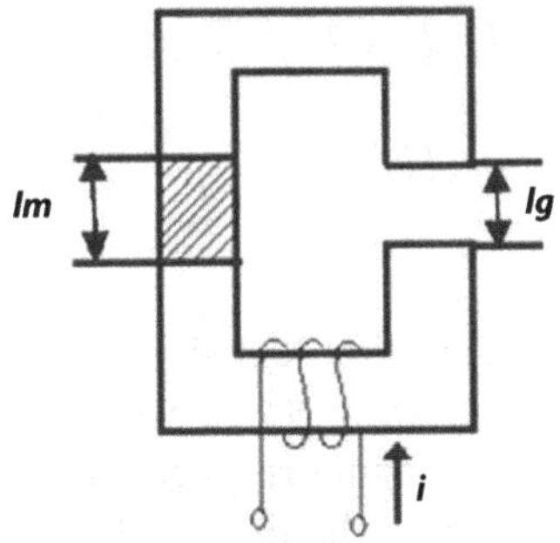

Figure 2.30 Electromagnetic device for example1.1.

Example 1.1.

The electromagnetic device showed in the Figure 2.30 has the following parameters:

No of turns of the coil, T_{ph} = 1200
Leakage factor, k= 1.25
Length of the magnet, l_m = 10mm
Length of the air gap, l_g = 0.5mm
Magnet cross sectional area = air gap cross section area

The magnet is of NdFeB type with a remanent flux density of 1.1T and coercivity of 0.86 x10^6 A/m is used. The demagnetisation characteristic of this magnet is linear in nature. Determine the air gap flux density when the coil carries 10 A current.

Soln:

$$\text{Magnet flux density}, B_m = B_r + \mu_0\mu_{rm}H_m \tag{2.46}$$

where B_r is the remanence flux density in Tesla, H_m is the magnet field intensity in A/m, μ_0 is the permeability of air and μ_{rm} is relative permeability of the magnet

At , $H_m = H_c$; $B_m = 0$

$$\therefore \mu_{rm} = \frac{B_r}{\mu_0 H_c} = 1.0062$$

Neglecting the mmf required by the iron part, the mmf produced by the coil will be exactly the same as the mmf required by the air gap and the mmf produced by the PM.

$$H_m l_m + H_g l_g = T_{ph} I$$

$$or, H_m = \frac{T_{ph} I - H_g l_g}{l_m} \tag{2.47}$$

Again, we know, air gap flux density, $B_g = \mu_0 H_g$, Air gap Flux, $\varphi_g = B_g A_g$ and magnet flux, $\varphi_m = B_m A_m$ where A_g and A_m are the cross sectional area of air gap and magnet respectively.

Considering the leakage flux, the magnet flux, $\varphi_m = k\varphi_g$ where k is the leakage factor.

$$\therefore \varphi_g = \frac{\varphi_m}{k}$$

$$or, B_g A_g = \frac{B_m A_m}{k}$$

$$or, B_g = \frac{1}{k}\frac{A_m}{A_g} B_m \tag{2.48}$$

Substituting the eqn 2.47 and 2.48 in to eqn 2.46 , the magnet flux density can be obtained as

$$B_m = \frac{B_r + \mu_0 \mu_{rm}\left(\frac{T_{ph} I}{l_m}\right)}{1 + \mu_{rm}\frac{l_g}{l_m}\frac{A_m}{A_g}\frac{1}{k}} = \frac{1.1 + 4\pi \times 10^{-6} \times 1.0062 \times \left(\frac{1200 \times 10}{10}\right)}{1 + 1.0062 \times \frac{0.5}{10} \times \frac{1}{1.25}}$$

$$= 1.072\ \text{T} \qquad \left[\begin{array}{c} As, A_m = A_g \\ \therefore \frac{A_m}{A_g} = 1 \end{array}\right]$$

$$\therefore B_g = \frac{1}{k}\frac{A_m}{A_g} B_m = \frac{1.072}{1.25}$$

$$= \mathbf{0.858T\ Ans}$$

Example 1.2

A 4 pole surface mount permanent magnet synchronous motor (SPMSM) has 36 stator slots. The slots are semi closed in shape having a opening of 2.2 mm.

The length of air gap along d- and q-axis are 4.4 mm and 5.4 mm respectively. The stator winding has 468 turns per phase. The inner stator bore diameter and length of the stator core are 82 mm and 102 mm respectively. The height of the SmCo magnet is given as 5 mm. Determine d-axis and q-axis armature reaction reactances. Consider the d-axis and q-axis armature reaction form factors as 0.966 and 1.014 respectively.

Soln: Given,
Stator bore dia, D = 82 mm,
Length of stator core, L=102mm,
Turns per phase =468
Length of air gap along d-axis, g = 4.4 mm
Length of air gap along q-axis, g_q = 5.4 mm
No of pole, P =4; No of slots, S_s = 36
Width of slot opening, W_s = 2.2 mm
d-axis armature reaction form factor, k_{fd} = 0.966
q-axis armature reaction form factor, k_{fq} = 1.014
Height of PM, h_m = 5 mm

Carter's coefficient, $k_C = \frac{2}{\pi}\left[tan^{-1}y - \frac{1}{\pi}log\sqrt{(1+y^2)}\right]$

Here, $y = \frac{W_s}{2g} = \frac{2.2}{2\times4.4} = 0.25$

$$\therefore k_C = \frac{2}{\pi}\left[tan^{-1}(0.25) - \frac{1}{\pi}log\sqrt{(1+0.25^2)}\right] = 0.155$$

Relative recoil permeability for SmCo = 1.1

Effective gap length along d-axis, $g' = gk_Ck_{sat} + \frac{h_m}{\mu_{rrec}} = 4.4\times0.07\times1 + \frac{5}{1.1}$

=4.85 mm

The pole pitch, $\tau = \frac{\pi D}{P} = \frac{\pi\times82}{4} mm = 64.4$ mm

The internal length L_i = 0.95x L =0.95x102 =96.9 mm

The d-axis armature reaction reactances, $X_{ad} = k_{fd}X_a = 4m\mu_0 f . \frac{(T_{ph}k_{w1})^2\tau L_i}{\pi pg'}k_{fd}$

$$= 4\times3\times4\pi\times10^{-7}\times50\times\frac{(468\times0.956)^2}{\pi\times2}\times\frac{0.0644\times0.0969}{4.85\times10^{-3}}\times0.966\Omega$$

$= \mathbf{14.9\ \Omega}$ **Ans**

Now, effective gap length along q-axis, $g_q' = g_q k_C k_{satq} = 5.4 \times 0.07 \times 1 =$ $0.378mm$

The q-axis armature reaction reactances, $X_{ad} = k_{fq} X_a = 4m\mu_0 f \frac{(T_{ph}k_{w1})^2 \tau L_i}{\pi p g_{q'}} k_{fq}$

$$= 4 \times 3 \times 4\pi \times 10^{-7} \times 50 \times \frac{(468 \times 0.955)^2}{\pi \times 2} \times \frac{0.0644 \times 0.0969}{0.378 \times 10^{-3}} \times 1.041\Omega$$

$= \mathbf{411.2\ \Omega}$ **Ans.**

Ex1.3: A 2 pole permanent magnet synchronous motor has 12 slots in its stator. Calculate the required skew angle to avoid the cogging torque.

Soln: Given, the no of stator slots is 12, hence no of cycles of cogging torque, N_{cl} for one revolution of rotor will be 12. As per eqn 1.43 the skew angle, $\theta_{sk} = \frac{360}{N_{c1}}$ degree electrical

$$= \frac{360}{12} \text{degree electrical}$$

$= 30$ degree electrical **Ans.**

References

[1] Kenjo T, Nagamori S, Permanent Magnet and Brushless DC motors: Clarendon Press, Oxford, 1985

[2] Parker RJ. Advances in Permanent Magnetism. New York: J Wiley and Sons,1990.

[3] GilbertW.DeMagnete,MagneticisqueCorporibusetdeMagnoMagneteTellure (On the Magnet, Magnetic Bodies and on the Great Magnet the Earth). London: 1600 (translated 1893 by Mottelay PF, Dover Books).

[4] Moosa I.J, History of Permanent Magnet, IJRDT, Volume: 2, Issue: 1, JULY-2014

[5] Heusler F, Stark W, Haupt E. Verh. Deutsche Phys., Gessel 5, 1903

[6] W. R. Lee, "Magnetism an Introductory Survey", (1970), Dover Publication, INC, NEW YORK.

[7] Parker RJ. Advances in Permanent Magnetism. New York: J Wiley and Sons, 1990.

[8] K. J. Strnat, G. Hoffer, J. Olson, W. Ostertag, J. J. Becker, J. Appl. Phys. 38 (1967) 1001.
[9] K. Atallah and D. Howe, Application of Halbach cylinders to brushless AC servo motors, IEEE Transactions on Magnetics, 34(4 pt 1), 2060–2062, 1998.
[10] M. Marinescu and N. Marinescu, Compensation of anisotropy effects in flux-confining permanent- magnet structures, IEEE Transactions on Magnetics, 25(5), 3899–3901, 1989.
[11] J. Ofori-Tenkorrang and J. H. Lang, A comparative analysis of torque production in Halbach and conventional surface mounted permanent-magnet synchronous motors, *IAS" 95. Conference Record of the 1995 IEEE Industry Applications Conference. ThirtiethIAS Annual Meeting (Cat. No. 95CH35862)*, pp. 657–663.
[12] R. F. Post and D. D. Ryutov, Inductrack: A simpler approach to magnetic levitation, *IEEE Transactions on Applied Superconductivity*, 10(1), 901–904, 2000.
[13] S. R. Holm, H. Polinder, J. A. Ferreira *et al.*, Comparison of three permanent magnet structures with respect to torque production by means of analytical field calculations, *International Conference on Power Electronics, Machines and Drives (IEE Conf. Publ.No. 487)*, pp. 409–414.
[14] S.-M. Jang, H.-W. Cho, S.-H. Lee *et al.*, The influence of magnetization pattern on the rotor losses of permanent magnet high-speed machines, *IEEE Transactions onMagnetics*, 40(4 II), 2062–2064, 2004.
[15] Z. Q. Zhu and D. Howe, Halbach permanent magnet machines and applications: A review, *IEE Proceedings—Electric Power Applications*, 148(4), 299–308, 2001.
[16] P. H. Mellor and R. Wrobel, Optimisation of a brushless motor excited by multi-polar permanent magnet array, *International Electric Machines and Drives Conference (IEEECat. No. 05EX1023C)*, pp. 649–654, 2005.
[17] . D. L. Trumper, M. E. Williams, and T. H. Nguyen, Magnet arrays for synchronous machines, *IAS'93. Conference Record of the 1993 IEEE Industry Applications ConferenceTwenty-Eighth IAS Annual Meeting (Cat. No. 93CH3366-2)*, pp. 9–18, 1993.
[18] T. R. Ni Mhiochain, J. M. D. Coey, D. L. Weaire *et al.*, Torque in nested Halbach cylinders, *IEEE Transactions on Magnetics*, 35(5 pt 2), pp. 3968–3970, 1999.
[19] Gieras J. F., Permanent Magnet Motor Technology: Design and Application, 3rd ed, CRC Press, Taylor and Francis Group, 2010
[20] Merrill FW. Permanent magnet excited synchronous motors. *AIEE Trans 72 Part III (June)*:581–585, 1953.
[21] Jahns TM, Kliman GB, Neumann TW. Interior PM synchronous motors for adjustable-speed drives. IEEE Trans on IA 22(4):738–747, 1986.

[22] J. F. Gieras, "Analytical approach to cogging torque calculation in PM brushless motors," *IEEE International Electric Machines and Drives Conference, 2003. IEMDC'03.*, Madison, WI, USA, 2003, pp. 815-819 vol.2

[23] A. Cassat, M. Williams, and D. MacLeod, Effect of skew on brushless dc motor exciting forces, *Proceedings of the Annual Symposium on Incremental Motion Control Systemand Device*, p. 148, 1993.

3

Hysteresis Motor

Richa Jain[1] and Sudha Gulati[2*]

[1]Department of Physics, Motilal Nehru College, New Delhi, India
[2]Department of Physics, Kalindi College, University of Delhi, East Patel Nagar, New Delhi, India

Abstract

A hysteresis motor is an alternating current motor having a rotating magnetic field. This motor is a self-starting synchronous motor that operates upon the hysteresis effect without DC excitation. Hysteresis motors have many applications due to their outstanding characteristics such as self-starting, constant speed, constant torque, noiseless operation, small start-up current, and low cost. The main five components of a hysteresis motor are the stator, single-phase stator winding, rotor, shaft, and shading coil. A rotor made entirely or in part of ferromagnetic material, and a stator that is properly coiled and stimulated to provide the elements of a spinning magnetic field The torque is produced by the interaction of magnetomotive force with the flux induced in the rotor, which transforms electrical energy into mechanical energy. A hysteresis motor's performance is mainly dependent on the composition of the rotor ring. The magnetic properties of the rotor material have a significant impact on the performance of the motor. The main types of hysteresis motors are cylindrical, circumferential-field, disk and axial-field hysteresis motors.

Analytical and finite element (FE) modeling is used to study these motors in terms of torque-current linearity and to enhance torque capability. This modeling is based on the normal magnetization curve of the hysteresis material. These motors have various applications such as sound-producing equipment, Sound recording instruments, timing devices, electric clocks, high-quality record players, and teleprinters. These motors have some drawbacks such as quite a low efficiency, low-power factor, long-starting time, and low torque density. Scientists are trying to find ways to fill this gap. Improvements in metallurgy have led to alloys with magnetic and physical characteristics easier to design hysteresis motors proficient for most effective operation than previously imagined. In the current work,

**Corresponding author*: sudhagulati@kalindi.du.ac.in

Jigneshkumar P. Desai (ed.) Special Electrical Machinery, (61–78) © 2024 Scrivener Publishing LLC

principle, construction, important performance-improving aspects, applications, and limitations of hysteresis motors are all thoroughly explored in the current work.

***Keywords*:** Hysteresis, model, stator, rotor, torque

3.1 Introduction

Electrical machines are considered the main technology for the advanced industrial revolution as Industry 4.0 and 5.0 which are based on smart machines and robots. To this end, the hysteresis motor could be a great choice due to its outstanding characteristics such as self-starting, constant speed, constant torque, noiseless operation, small start-up current, and low cost [1]. The hysteresis motor is an alternating current (ac) motor with a rotating magnetic field. This motor is a self-starting synchronous motor that operates upon the hysteresis effect without DC excitation. It was first explained by Steinmetz in 1917. In this motor, the driven torque depends on magnetic hysteresis [2].

The hysteresis motor is a type of synchronous motor that operates based on the hysteresis phenomenon in magnetic materials. It uses a rotor made of a magnetic material that has a high coercive force, which means that it is difficult to magnetize or demagnetize. The stator of the motor has a rotating magnetic field that induces a magnetic field in the rotor. Due to the hysteresis effect, the rotor lags the magnetic field of the stator, which causes it to rotate.

Hysteresis motors are known for their smooth and quiet operation, and they are often used in low-power applications, such as clocks, turntables, and other precision instruments. They are also used in medical equipment, where low noise and vibration are important [3].

The main components of this motor are a stator, single-phase stator winding, rotor, shaft, bearings, and shading coil. The hysteresis motors are of many types such as cylindrical hysteresis, disk hysteresis, circumferential-field hysteresis, and axial-field hysteresis, motors. These motors have low power densities and power factors compared to others. The hysteresis motor is a self-starting synchronous machine having a simple configuration [4]. Its rotor is typically made up of semi-hard permanent magnet material and the stator is like the induction motor's stator. Mainly, the commercial hysteresis motors are of circumferential flux type. The design of the hysteresis motor depends on the choice of the hysteresis material of the rotor, stator poles, winding configuration, air gap dimensions, stator lamination geometry, etc [5].

The stator is the stationary part of the motor, and it consists of a laminated core made of magnetic material. The stator has windings that are arranged in a specific pattern to generate a rotating magnetic field. The rotor, on the other hand, is the rotating part of the motor and is made of magnetic material with high coercive force, such as chrome steel. Due to the hysteresis effect, the rotor lags the rotating magnetic field of the stator, which causes it to rotate.

The bearings are the components that support the rotor and allow it to rotate smoothly and quietly. They are often made of high-quality materials, such as ceramic, to minimize friction and wear [6].

There are four main classes of hysteresis motors: cylindrical disk, Disk, Circumferential-field, and Axial-field hysteresis motor. Disc-type hysteresis motors are appropriate for high-speed applications due to their firm structure, high power density and low axial length. Multistage axial field hysteresis motor has outstanding properties such as compactness, appropriate shape, robustness, and electric characteristics and could have many applications [7].

A hysteresis motor initially operates as a single-phase induction motor, but as it runs, it transforms into a synchronous motor. Its performance mainly depends on the material of the rotor ring. Depending on the kind of applications, the rotor ring can be manufactured using Alnico, Fe-Cr, Fe-36% Cobalt and Fe- Cr-Ni-Mo-C etc. Hysteresis loops with large area has better output power density, require high conductivity to enhance the initial torque, require the operation of the motor in normal loop for maximum efficiency and undergo a special heat treatment and hardening procedure [8].

Hysteresis motors have some limitations such as relatively low efficiency, low-power factor, low torque density, and long-starting time. These motors have many applications due to their outstanding characteristics such as self-starting, constant speed, constant torque, noiseless operation, and less start-up current. They can be utilized in high-speed applications due to which their core losses are the main portion of the total losses of these motors.

Although at present, hysteresis motors have very limited applications but their fundamental features could make them ideal for next-generation smart machinery. These motors show nonlinear characteristics and their behavior highly depends on the initial magnetization of the rotor active part. It is a challenging task to model, design and control the hysteresis motors as these motors have nonlinear nature. Many reliable models are proposed by researchers to study and find the torque capability theoretically and experimentally [9].

In the dynamic model, a circumferential-flux hysteresis motor has been constructed and stimulated. This model mainly depends on the properties of the B-H loop [10]. In the Finite element (FE) model, a magneto-static FE simulation is performed by the normal magnetization curve of the hysteresis material, and after that, the average torque is computed using a post-processing analysis. The simulated results were compared with experimental results [11].

Hysteresis motors can have applications where constant and low ripple torque is required such as electric clocks, timing devices, gyroscopes, audio equipment, magnetic ribbons, self-bearing motors, etc.

3.2 Construction

Steinmetz made an interesting group of self-starting multiphase synchronous motors called hysteresis motors in 1900 [12]. The hysteresis motor is a rotating magnetic field type of ac motor as shown in Figure 3.1. This motor has a uniform air gap and no DC excitation and uses the hysteresis properties of magnetic materials to generate the torque. It is often used in synchronous motors that need a smooth, soft torque and a simple design with standard three-phase stator windings.

A hysteresis motor consists of a stator, the single-phase stator winding, the rotor, the shaft, and the shading coil as shown in Figure 3.2. The stator

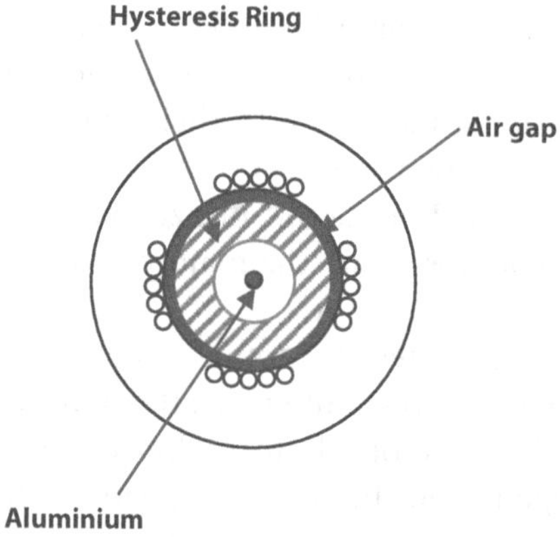

Figure 3.1 Outline of hysteresis motors.

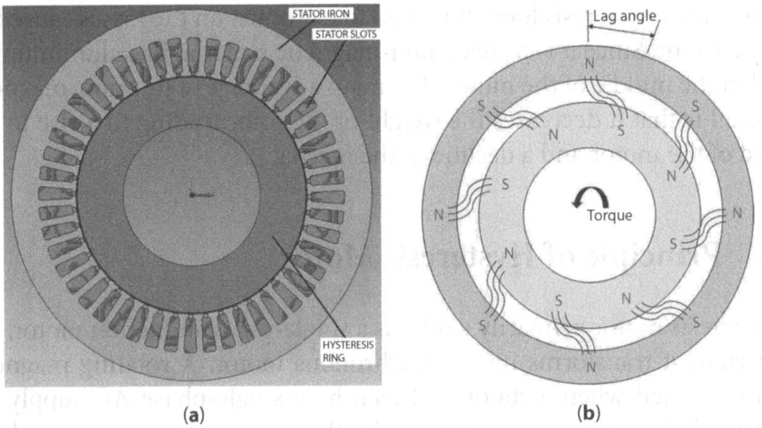

Figure 3.2 (a) Typical Hysteresis motor; (b) torque production principle. Reproduced from ref. 15 under a Creative Commons Attribution 3.0.

and rotor are the two most important parts of a hysteresis motor. The rotor of a hysteresis motor is just a solid ring made of a hysteresis material. The stator of a hysteresis motor is wound like in a regular motor.

The stator is wound with both main and secondary coils so that a rotating magnetic field can be produced. In some types of hysteresis motors, it can also be a shaded pole type. The stator of a hysteresis motor is made so that a single-phase supply can be used to create a synchronized rotating field. Stator coil properties include stator coil resistance and stator coil leakage inductance

Stator Coil Resistance (R_s) - The total (DC) resistance of the stator coils for a single phase can be easily measured directly by using an ohm-meter in the laboratory. An equation for R_s is given by Darabi *et al.* in terms of the number of windings and wire gauges [13].

Stator Coil Leakage Inductance (L_{ls}) - The leakage inductance L_{ls} can be expressed in terms of leakage reactance X_1

$$L_{ls} = \frac{X_1}{w_b}$$

Where w_b is the angular frequency of polyphase input. Darabi *et al.* have reported motor dimensions that can be substituted into their equations [10].

The rotor is made of hysteresis material which has several hysteresis rings made of chrome steel or alnico. The rotor does not have teeth or winding.

It has a large hysteresis loop. It is used to cut down on the losses caused by eddy current. Since it is heavier, non-magnetic material like aluminium, is used in the middle of the motor. The main advantage of this non-magnetic material is that it decreases the weight of the rotor leading to a rise in the speed of the motor and a decline in the inertia.

3.3 Principle of Hysteresis Motor

A hysteresis motor primarily works as a single-phase induction motor, but as it runs, it transforms into a synchronous motor. A rotating magnetic field is created when a stator is driven by a single-phase AC supply. To withstand the revolving magnetic field, the main and auxiliary windings can be constantly supplied with supply at the beginning and during running operations. This field causes the rotor to produce poles.

Due to the rotor material, there is dominance of hysteresis and the rotor pole axis lags behind the axis of the revolving magnetic field. This causes rotor poles to become attracted to the moving stator field poles. Consequently, the rotor is subjected to a torque known as hysteresis torque. Torque remains constant regardless of speed. While the stator field axis goes forward, the rotor pole strength is maintained due to its strong retentivity. Hence, when retentivity increases, hysteresis torque increases.

Both hysteresis torque and torque due to induced eddy currents in the rotor cause the rotor to initially begin rotating. The stator pushes the rotor into synchronism whenever the speed is close to that of synchronous motion. In this scenario, the torque caused by eddy currents also disappears when the relative motion between the stator field and rotor diminishes. All that is present is hysteresis torque, which keeps the rotor spinning at synchronous speed. The constant magnetic locking between the stator and rotor is ensured by the strong retentivity. The Hysteresis Motor either rotates at synchronous speed or not at all because of the magnetic locking concept.

Under ideal steady-state conditions, the magnetic poles on the rotor lag the stator field by an angle δ that grows with load torque even though they rotate at the same speed as the stator field. The lag angle is zero when the load torque is zero. It equals some maximum value δ_m at the maximum load torque prior to dropping below synchronous conditions. The hysteresis Ellipse Parameters of Rotor Material B_m, μ, and δ_m define the maximal area hysteresis ellipse

$$H = \frac{B_m}{\mu}\cos(\theta)$$

$$B = B_m \cos(\theta - \delta_m)$$

where θ is variable, μ is permeability, and δ is the angle by which the B-field lags the H-field, which is rotating is generated by the stator coils. B_m is the maximum value of B on the hysteresis loop. These variables are influenced by the past history, particularly the peak value H_m ($=B_m/\mu$) of the field of the rotor. On a rotor material sample, a sinusoidally fluctuating field can be used to experimentally evaluate the parameters. The lowest harmonic ($\theta = \omega t$) is found in the measured resulting waveform for B for different values of H to find δ_m. It has values between 30 and 50 [13].

3.4 Classes of Hysteresis Motor

Four main classes of hysteresis motors are cylindrical, Disk, circumferential-field, and Axial-field hysteresis motor.

Cylindrical hysteresis motors - The cylindrical rotor of cylindrical hysteresis motors can be entirely formed of hysteretic material or alternatively can have an inner core made of non-hysteretic material, such as a portion of the shaft unit. The stator encloses the rotor, with an air gap between the stator's inner surface and its inner radial surface known as the rotor stator air gap.

Rotor Gap Inductance L_m can be expressed as

$$L_m = \frac{2mk_w^2 N_w^2 \mu_0 l \mu}{\pi p^2 \ell_g}$$

where m is the number of phases, p is the number of pole pairs, k_w is the winding coefficient, N_w is the number of stator windings per phase, r_g is the mean radius of the air gap, ℓ is the axial thickness of rotor, ℓ_g is air gap width, μ_o is the permeability of free space and μ is the permeability of magnetic material. Rotor Equivalent Resistance can be expressed as

$$R_h = \frac{w_b k_w^2 N_w^2 \mu V_r \sin(\delta)}{\pi^2 r_r^2}$$

where r_r is the mean radius of the hysteretic ring of the rotor and V_r ($= 2\pi r_r t_r \ell$) is the volume of the rotor cylinder.

Disk hysteresis motors - Rotors in disc hysteresis motors have an annular ring shape. The stator and ring's flat surface are separated by an air gap. The ring is frequently fastened to a centrally located disk-shaped support that is perpendicular to a shaft [14].

Axial-field hysteresis motor - The ring support of an axial-field hysteresis motor is built of a magnetic substance with practically infinite magnetic permeability. The name of this motor is derived from the fact that the disc is thin and the field lines in the rotor are axial to the disc before the lines pass through the support material. If r_r is the mean radius of the rotor ring, $\ell = r_o - r_i$, r_o is outer radius of ring and r_i is inner radius of ring, V_r ($=2\pi r_r \ell t_r$)is volume of rotor ring, t_r is thickness of rotor ring, then rotor gap inductance (L_m), rotor equivalent resistance for hysteresis (R_h) and rotor inductance (L_{hr}) are given by

$$L_m = \frac{2mk_w^2 N_w^2 \mu_0 r_r \ell}{\pi p^2 \ell_g}$$

$$R_h = \frac{w_h m k_w^2 N_w^2 V_r \mu \sin(\delta)}{\pi^2 p^2 t_r^2}$$

$$L_{hr} = \frac{m k_w^2 N_w^2 V_r \mu \cos(\delta)}{\pi^2 p^2 t_r^2}$$

Circumferential-field hysteresis motor - A ring support for a circumferential-field hysteresis motor is built of a non-magnetic substance whose magnetic permeability is taken to be zero. The name comes from the fact that the magnetic field lines in the rotor will primarily surround the ring.

3.5 Hysteresis Torque and Hysteresis Loss Equation

Consider a hysteresis motor having supply frequency f, eddy current power loss P_E, eddy current frequency f_E, flux density B. The expression of P_E can be expressed as

$$P_E = k_E f_E^2 B^2 \tag{3.1}$$

Where k_E is eddy current constant.

The expression of f_E in terms of slip s (= difference between the rotor speed and the synchronous speed divided by the synchronous speed) could be written as

$$f_E = s\,f \tag{3.2}$$

Using eq. (3.2), eq. (3.1) becomes

$$P_E = k_E s^2 f^2 B^2 \tag{3.3}$$

The force that twists or rotates an object into rotation is referred to as torque. Consider example of a door. It is obvious that the door cannot rotate unless a force is applied. But not every force is able to carry out this task. For instance, a force applied to the door's hinge line won't result in any rotation at all. In contrast, the most efficient way to cause rotation of the door is to apply a force of a particular size at a right angle to the door at its outer edge. It does so because the torque generated by this force causes the hinge to rotate. It takes more than just force to rotate an object; it's also vital to consider where and how the force is applied. Power and torque have a straight direct relationship with one another. The scalar product of torque and angular velocity can be used to numerically express the power of a rotating object. The rotor experience torque τ_E due to eddy current and is given by (Figure 3.2(b))

$$\begin{aligned} \tau_E &= \frac{P_E}{s w_s} \\ &= \frac{k_E f^2 B^2}{w_s} s \\ &\propto s \qquad \text{(all other parameters are constant)} \end{aligned} \tag{3.4}$$

The above equation indicates that rotor rotates with synchronous speed, the slip s becomes zero leading to vanishing of eddy current component.

The hysteresis loss can be expressed as

$$P_E = k_H B^{1.6} f_E$$
$$= k_H B^{1.6} sf \qquad \text{(using eq. (3.2)) (3.5)}$$

Where k_H is a constant.
The corresponding torque can be expressed as

$$\tau_E = \frac{P_H}{sw_s}$$
$$= \frac{k_H f B^{1.6}}{w_s}$$

The above equation shows that the hysteresis component is constant for all the speeds of rotor. In the rotor of hysteresis motor, hysteresis loss is produced which depends on the area of hysteresis loop. The heat is dissipated in the rotor due to the hysteresis loss. The heat energy dissipated per minute (W) in the rotor in terms of hysteresis loss per revolution (E_H) and N_s revolutions per minute can be written as

$$W = N_S E_H$$

The corresponding power is

$$P_S = \frac{W}{t}$$
$$= \frac{N_S E_H}{t} \qquad (3.6)$$

The mechanical power used for driving the rotor is the source of power dissipated in the rotor and is given by

$$P_H = \frac{2\pi N_s}{w} \tau_H \tag{3.7}$$

where, τ_H is the torque exerted on the rotor [10].

Equating eq. (3.9) and (3.10) gives

$$\frac{2\pi N_s \tau_H}{60} = \frac{N_s E_H}{t}$$

Thus $\tau_H = \dfrac{E_H}{2\pi}$

Torque Speed characteristic of Hysteresis Motor is shown in Figure 3.3.

Curve 'a' is the ideal curve whereas the curve 'b' is the practical hysteresis motor curve. The hysteresis motor shows a different torque-speed characteristic as compared to an induction motor. The torque of an induction motor decreases to zero at synchronous speed, however that of hysteresis motor remains constant at all speeds, including synchronous speed. As it runs at a steady speed and has a smooth rotor, the hysteresis motor has a far lower noise level than an induction motor. The best sound reproduction equipment, such as record players and tape recorders, employ this kind of motor because it runs quietly and smoothly. Electric clocks and other timing devices also use it. Figure 3.4 shows flowchart of finite element based method [15].

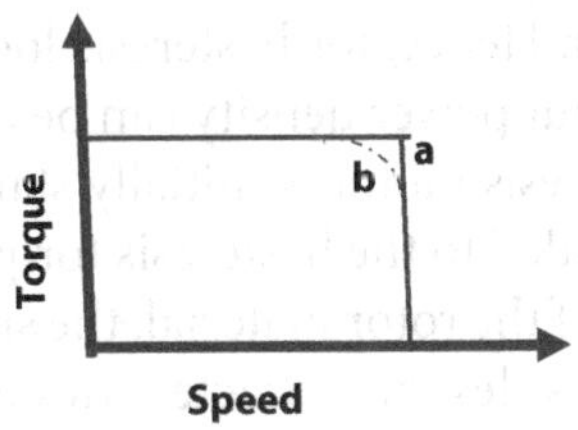

Figure 3.3 Variation of torque with speed.

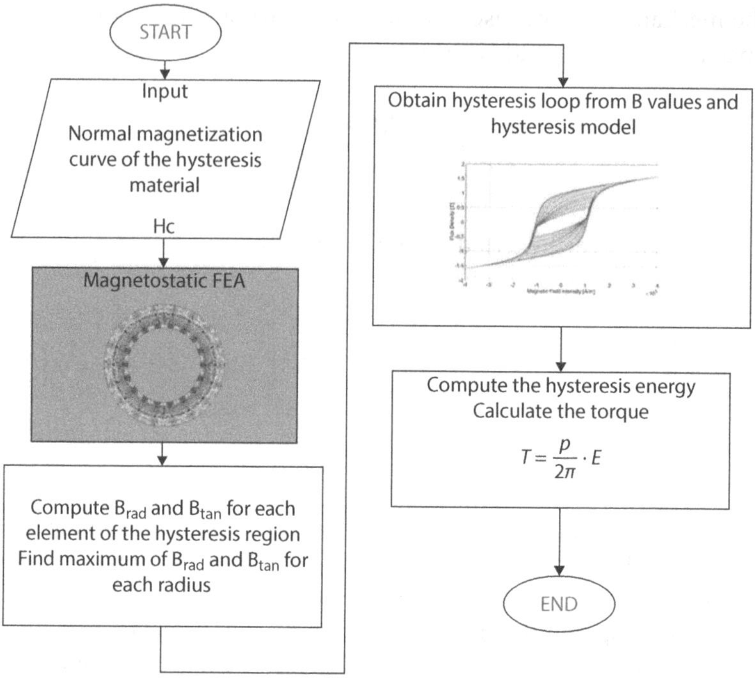

Figure 3.4 Flowchart of the FE model [15].

3.6 Selection of the Appropriate Material for Hysteresis Motor

A hysteresis motor's performance is solely dependent on the material of the rotor ring, which is selected after taking into account the following factors:

- The area of the hysteresis loop and the volume of the rotor ring are directly proportional to the output power of the hysteresis motor. Hence, for hysteresis loops having a larger area, better output power density can be achieved.
- When the hysteresis motor is initially started, the eddy current torque is added to the hysteresis torque. The conductivity and volume of the rotor material, the size of the hysteresis loop, and various design characteristics all affect eddy current torque. So, choosing a suitable material with a higher conductivity can increase the initial torque in the hysteresis motor.

- If the magnetic field intensity of the hysteresis loop is low and the loop's permeability is high, the stator magnetising current and, as a result, copper loss, will be minimal. Thus, the machine operates at its maximum efficiency when hysteresis motor operates on the the normal loop (one of the hysteresis loops).
- The hysteresis materials typically undergo a special heat treatment and hardening procedure. Hysteresis materials become mechanically hard and hence pose certain manufacturing challenges. This issue should be taken into consideration when selecting the hysteresis material.
- For a hysteresis motor, higher output torques can be obtained in more hysteresis delay angles.
- MMF harmonic content, including slotting effects, causes minor loops and related parasitic losses of the rotor rings [15, 16]. The slope of minor loops is understood to be equal to the slope of the magnetising curve at the maximal point of a hysteresis loop. The minor loops slope and the parasitic losses are inversely correlated. The value of the magnetising curve slope at the greatest point of the normal loop is therefore one of the crucial considerations in choosing the rotor material.

3.7 Models

Although hysteresis motors have limited applications till date due to their low efficiency, low-power factor, long-starting time, and low torque density etc. but it could be a substitution for high-speed motors in future. At high speed, a rotating electrical machine may have less efficiencies, stability at higher speeds, overheating etc. These drawbacks can be removed in hysteresis motors [15].

The solid-rotor hysteresis motor has many advantages such as:

(i) High mechanical strength so that it works properly at high peripheral speed.
(ii) Limit the rotor dynamic instability difficulties which occur due to higher flexural stiffness and critical speeds.
(iii) Hysteresis motors are less likely to overheat the rotor because of their synchronization Feature.

(iv) In these motors, Curie temperatures of magnetic materials are higher than that of permanent magnets.
(v) These motors have ability to improve their torque performance of rotor.

It is a challenge to model, design and control the hysteresis motors because of their nonlinear nature. Therefore, various models are suggested by researchers for proper functioning of this motor.

3.7.1 Dynamic Model

In the dynamic model, a circumferential-flux hysteresis motor has been established and stimulated. A linear time-varying equivalent circuit is recommended for the rotor as a series inductive impedance and dependent voltage source as shown in Figure 3.5. It depends on the changes of load torque and stator voltage. The main advantage of this model is to study the properties of the B-H loop. The hysteresis lag angle is determined using the applied stator voltage and available load torque. It does not depend on the frequency of rotor but based on the shape of the hysteresis loop [17].

There is a close agreement between simulated and measured results. This model could be used for fault diagnosis, sensor less control, closed-loop vector control and the identification of parameters.

3.7.2 Finite Element Model

In the Finite Element model, the hysteresis torque is not stimulated using FEA software instead it is computed using Steinmetz equations for laminated steels. Hysteresis torque cannot be found using this model directly and therefore a post-processing hysteresis model is needed. In FE model,

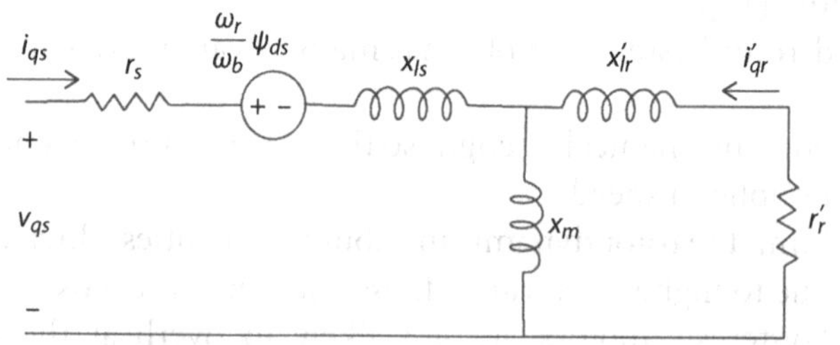

Figure 3.5 A dynamic model hysteresis motor. Reproduced from ref. 14 under a Creative Commons Attribution.

the first harmonic of the magnetomotive waveform at the surface of the rotor ring is used to determine the hysteresis losses of the rotor and electromagnetic torque with a model [18].

The FE model is analysed using with initial permeability and initial remanence. The tangential component of the flux density (B_{tan}) is determined for each ring element and evaluation of the maximum value of B_{tan} is found. Hysteresis model is employed to know the actual value of magnetic field. New inputs are calculated for the finite element analysis.

3.8 Limitations

Magnetic hysteresis machines feature a straightforward design with a laminated or solid rotor constructed of a semi-hard magnetic material and a standard multiphase stator winding. Hysteresis machines are described as self-starting motors in the literature, capable of accelerating and synchronising high-inertia loads, producing consistent torque from a halt to synchronised speed [19]. Furthermore, controlled clutches and couplers can make use of magnetic hysteresis [20]. Hysteresis machines have low power factors and limited power efficiency when compared to other motor technologies. Because of this, the range of their uses in electric machinery is quite limited. They include gyroscopes, clocks, magnetic ribbons in recorders, self-bearing motors, and other low-power precision machinery [19, 21, 22]. Hysteresis machines, however, are in a hopeful situation because of the recent rise in demand for high-speed motors. The following basic problems with rotating electric machinery occur at high speeds

Mechanical strength - The highest achievable peripheral speed of a rotor is proportional to the tensile strength of its material, making mechanical strength a significant restriction. This hinders technologies that rely on sintered permanent magnets or laminated arrangements. To partially alleviate this issue, sleeves and other retaining features can be used, however doing so will lower the efficiency of the machine.

Rotordynamics - Because a rotor may cross one or more critical speeds during speed transitions, rotordynamics is a crucial aspect. Solid rotors have a high rigidity that can eventually help to drive crucial speeds over the machine's working limits.

Thermal behavior - When rotor eddy currents are used to generate torque in motors, thermal behaviour is a prevalent issue. Even though permanent magnet synchronous motors are less likely to experience rotor overheating, their materials are limited to relatively severe maximum temperature ranges because of the possibility of demagnetization.

3.9 Future Prospects

Hysteresis motors have been used in a variety of applications over the years, particularly in situations where precise control of speed and torque is required. However, due to the emergence of more efficient and cost-effective motor technologies, such as permanent magnet synchronous motors and brushless DC motors, the use of hysteresis motors has declined. Despite this, hysteresis motors still have some unique advantages that may make them a viable option for certain niche applications. For example, they can operate with very low torque ripple, making them suitable for precision applications such as robotics and medical equipment. They are also very quiet and do not produce much electromagnetic interference, which is important in certain industries such as broadcasting and telecommunications. There has been ongoing research in recent years aimed at improving the performance and efficiency of hysteresis motors, particularly through the use of advanced materials and improved control algorithms.

One area of interest has been the development of high-temperature superconducting materials for use in hysteresis motors, which could potentially improve their efficiency and power density [23].

There have been publications in the literature on the investigation of the magnetic properties of a steel alloy (Fe-Cr-Ni-Mo-C) for use in the rotors of hysteresis motors. These characteristics were compared with those of a few materials that are frequently used to make the rotors for hysteresis motors. This comparison demonstrated how using this alloy with a suitable design can raise the hysteresis motor's efficiency and power factor. Although the recommended material has a lower power density than the other materials, it might still be appropriate for some applications, such as gyroscopes, where a higher rotor moment inertia is helpful. Also, other hysteresis materials include rare-earth elements and expensive components like cobalt while alloy steel is affordable and readily available, therefore this alloy was suggested for use in hysteresis machines [24].

In order to operate in the liquid nitrogen medium, new types of electrical machines with bulk YBCO (yttrium barium copper oxide) and plate-shaped Bi-2223/Ag elements were built and tested. In comparison to traditional motors, these motors had a lower specific weight and a higher output power. In the near future, electrical machines incorporating bulk hysteresis elements could be used in a variety of applications like aerospace technology, transport systems etc. [25].

Modeling, designing, and controlling hysteresis motors is a complex endeavour due to their nonlinear behaviour. Current research efforts have focused on developing new options in this environment [26].

References

1. Tsai, W.-H. Green Production Planning and Control for the Textile Industry by Using Mathematical Programming and Industry 4.0 Techniques. Energies 11, 2072, 2018.
2. Tahanian, H., Darabi, A., New design method for multi-stack disc-type hysteresis motors based on analytical calculations. IET Electr. Power Appl., 12 , 2, 222-230, 2018.
3. L. Zhou and D. L. Trumper, Magnetically Levitated Linear Stage With Linear Bearingless Slice Hysteresis Motors. *IEEE/ASME Transactions on Mechatronics*, 26, 1084-1094, 2021.
4. Bertotti, G. Hysteresis in Magnetism: For Physicists, Materials Scientists, and Engineers; Academic: New York, NY, USA, 433–504, 1998.
5. L. Zhou, W. Gruber, and D. L. Trumper, Position Control for Hysteresis Motors: Transient-Time Model and Field-Oriented Control. *IEEE Transactions on Industry Applications*, 54, no. 4, 3197-3207, 2018.
6. Xuehui Gao, Adaptive Neural Control for Hysteresis Motor Driving Servo System with Bouc-Wen Model. Complexity, 2018, 9765861, 2018.
7. Ghanbari, T., Sanati-Moghadam, M., Darabi, A.: 'Comparison between coreless and slotless kinds of dual rotor discs hysteresis motors', *IET Electr. Power Appl.*, 10, 133–140, 2016.
8. Kubota, T., Wakui, G., and Itagaki, M., "Hysteresis motor using magnetically anisotropic Fe-Cr-Co magnet," in *IEEE Transactions on Magnetics*, 34, 3888-3896, 1998.
9. Aishwarya Dhanagare, A.G.Thosar, Modeling of Hysteresis Motor, International Journal of Engineering Research and Development. 12, 5, 76-83, 2016.
10. Niasar, A. H., A novel time varying dynamic modeling for hysteresis Motor. *Scientia Iranica D*, 24, 1395-1409, 2017.
11. Jagiela, M., Garbiec, T. and Kowol, M., "Design of High-Speed Hybrid Hysteresis Motor Rotor Using Finite Element Model and Decision Process," in *IEEE Transactions on Magnetics*, 50, 861-864, 2014.
12. Steinmetz, C.P., The hysteresis motor, in Wolcott, T. (Ed.): 'Theory and calculation of alternating current phenomena', 293–296, McGraw, 1900, 3rd edn.
13. Darabi, A., Lesani, Ghanbari, H. T., Akhavanhejazi, A., (2007a), Modeling and optimum design of disk-type hysteresis motors, *Proc. Inter. Conf. Elect. Mach. and Sys.*, Seoul, Korea, 998, 2007.
14. Ishikawa, T., and Kataoka, T., Basic analysis of disc-type hysteresis motors, *Elec. Engng. Japan*, 101, 6, 659, 1981.
15. Rahman, M. A. Copeland, and Slemon, G. R., An analysis of the hysteresis motor III-Parasitic losses, *IEEE Trans. Power App. Syst*, PAS-88, 954, 1969.
16. Wakui, G. and Tomita, M., Effect of harmonic flux density on the characteristics of hysteresis motor, *Trans.IEE Jpn.*, 102-B(1), 25, 1982.

17. Niasar,H., Zare, M., and H. Moghbelli, H., Dynamic Modeling and Simulation of a Super-High-Speed Circumferential-Flux Hysteresis Motor. *Journal of Engineering*, 2013, 7, 2013.
18. Gallicchio, G.; Palmieri, M.; Di Nardo, M.; Cupertino, F. Fast Torque Computation of Hysteresis Motors and Clutches Using Magneto-static Finite Element Simulation. *Energies,12,* 3311, 2019.
19. Inácio, D., Inácio, S., Pina, J., *et al.*, Numerical and experimental comparison of electromechanical properties and efficiency of HTS and ferromagnetic hysteresis motors, *Journal of Physics: Conference Series*, 97, 12218, 2008.
20. Canova A., and Cavalli, F., Design Procedure for Hysteresis Couplers, *IEEE Trans. Mag.*, 44, 2381, 2008.
21. Rajagopal, K. R., Design of a Compact Hysteresis Motor Used in a Gyroscope, *IEEE Trans. Magn.*, 39, 3013, 2003.
22. Zhou, L., Imani Nejad,M., and Trumper, D. L., One-axis hysteresis motor driven magnetically suspended reaction sphere, *Mechatronics*, 42, 69, 2017.
23. Das, J. and Ray, R.N., Design and Modeling of Hysteresis Motor with High Temperature Superconducting Material in the Rotor using Finite Element Method, *Innovative Systems Design and Engineering*, 3, 6, 2012.
24. Ghanbari, T., Darabi, A. and Moghadam, M.S., Hysteresis Motor Using Heat Treated Fe-Cr-Ni-Mo-C Steel Alloy. *Journal of Electrical Systems*, 11, 49, 2015.
25. Kovalev, L.K., Ilushin, K.V., Kovalev, K.L., Penkin, V.T., Poltavets, V.N., Koneev, S.A., Akimov, I.I., Gawalek, W., Oswald, B. and Krabbes, G., 2003. High output power electric motors with bulk HTS elements. *Physica C: superconductivity*, 386, 419, 2003.
26. Galluzzi, R., Amati, N. and Tonoli, A., 2019. Modeling, design, and validation of magnetic hysteresis motors. *IEEE Transactions on Industrial Electronics, 67,* 1171. 2019.

4

Switch Reluctance Motor

V.B. Babaria[1] and Jigneshkumar P. Desai[2*]

[1]*L.D. Engineering College, Ahmedabad, India*
[2]*U V Patel College of Engineering-Ganpat University, Mehsana, India*

Abstract
This chapter gives a general overview of the Switched Reluctance Motor (SRM) technology, including information on its background, design, use, and control strategies, as well as its advantages and disadvantages. The link between current, inductance, and torque output is highlighted by the development of the torque equation. The chapter goes on to examine several SRM converter topologies and their advantages and disadvantages. Elevators, washing machines, electric cars, robotics, and other devices are among the things that use SRM. The potential for high RPM and high-power applications, such as electric vehicles and traction systems, is part of the future scope of SRM technology. Modern computer software, AI, and finite element analysis developments have improved the design capabilities of SRMs, and commercial SRMs with up to 200 KW ratings have been created.

Keywords: Acoustic noise, dwell, harmonics, torque ripple

4.1 Introduction

SRM boasts a rich historical lineage, with its origins dating back to the early 19th century. Although the initial development of SRM took place in 1836, it faced challenges in terms of controllability, leading to limited popularity at that time.

However, a significant turning point occurred in the 20th century with the advent of electronics. The progression of controllers gained momentum, and by the 1980s, the integration of microprocessors, computers, and advanced power electronic switching circuits, along with the introduction

**Corresponding author*: jpd.fetr@gmail.com

Jigneshkumar P. Desai (ed.) Special Electrical Machinery, (79–98) © 2024 Scrivener Publishing LLC

of highly precise sensors, catalysed a wide array of applications for SRM. This motor found favor due to its adaptability to diverse power, speed, and torque demands. Notably, one of its distinguishing features is the absence of permanent magnets on the rotor and the elimination of wound stators, characteristics in contrast to conventional induction motors. This unique design translates to cost-effective manufacturing, heightened efficiency, and low operating temperatures—thereby preventing temperature-related issues associated with magnets. Additionally, the cooling process is simplified.

Moreover, these motors possess linear motor attributes, extending their applicability to an even broader spectrum of uses. The remarkable torque-to-weight ratio further amplifies their utility.

4.2 Construction of SRM

The SRM (Switched Reluctance Motor) operates through electromagnetic and electrodynamic principles, deriving its torque from the rotor's inclination to move towards positions that maximize the inductance of the energized winding. Figure 4.1 shows the construction of SRM motor along with flux distribution pattern. The motor employs ferro silicon sheets as its core material. Constructed with precision, both the stator and rotor are assemblies of insulated laminated stampings, preventing electrical interconnection via varnish insulation. These components feature projected poles for enhanced performance.

To avert the issue of cogging, the stator and rotor poles are intentionally designed to be dissimilar, ensuring they don't align perfectly. This divergence in pole configuration is demonstrated by various combinations such as 8 poles on the stator matched with 4 on the rotor, 4 stator poles with 2 rotor poles, 8 stator poles with 6 rotor poles, and so forth.

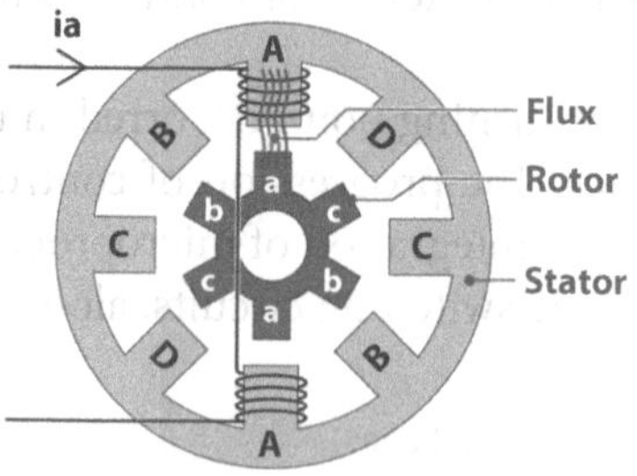

Figure 4.1 Rotor pole aligned with stator pole result in La maximum.

Uniformity is maintained among the stator poles, each equipped with identical coils. Counterpart coils on opposite poles are interconnected in a series, combining to form a phase. This arrangement results in a configuration where, for instance, 4 stator poles constitute 2 phases, 6 stator poles yield 3 phases, and 8 stator poles engender 4 phases. This pattern continues with increasing numbers of stator poles.

4.3 Operation and Characteristic of SRM

This motor operates with a fixed power supply, facilitated by a converter, a controller, power switching circuits, and shaft-mounted sensors. Current transformers are employed on the current-carrying phases. When the rotor poles align precisely with the stator poles, the field coil exhibits maximum inductance, represented by the formula L = N*2/S, with S representing the reluctance of the magnetic path. Notably, the flux path predominantly consists of iron, contributing negligible reluctance, and the path across the air gap is also quite short as depicted in Figure 4.1.

Conversely, when the rotor poles deviate from alignment with the stator poles, the inductance reaches its minimum point. During rotor rotation, a torque resulting from variable reluctance emerges, further contributing to the overall torque. This mechanism results in a flux distribution across the machine, which can be effectively analyzed using Finite Element Analysis (FEA) through computer techniques. This analysis aids in optimizing the machine's design and performance by providing insights into the distribution of magnetic flux.

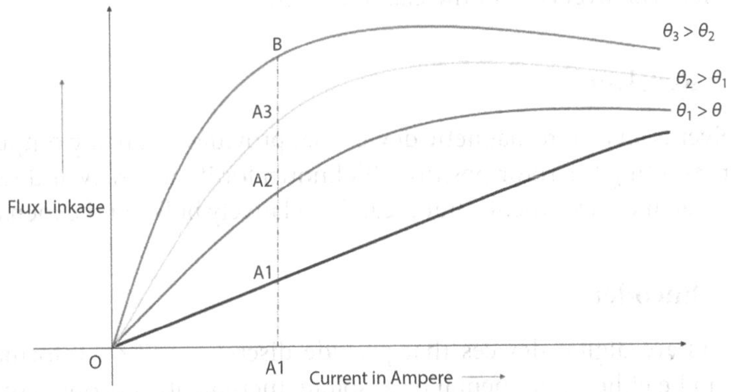

Figure 4.2 Flux linkage vs current plot of SRM.

Slope of this characteristic are shown Figure 4.2. Area OABC gives energy stored in induce system of motor. O-A-A_1-C_1 co-energy and OAC is field energy. As rotor rotates, co-energy change from OAA to OABO. This change of energy to with respect to time give torque.

$$\tau = \frac{dw_c}{d\theta} = \frac{OABO - OAA_1O}{\theta - \theta_3}$$
$$= \frac{OA_1BO}{\theta - \theta_3}$$

At high-speed back emf is offered by winding. Hence torque reduces, as inductance and stored energy reduces. Torque is high at low speed. There is no effect of polarity of current on torque. If $\frac{dL}{dt}$ is positive, then the motoring action and negative then generating action. Here torque is not uniform as rotor rotates and torque ripples are there. Distortion is introduced in current due to this. This generates harmonies in current. To filter this capacitor is required on AC side. This improves power factors also [1]. Hence size of supply KVA is not rising.

4.4 Classification of Rotor Position Sensor

The classification of rotor position sensors for Switched Reluctance Motors (SRMs) involves categorizing the various types of sensors used to determine the exact position of the rotor within the motor. Accurate rotor position sensing is crucial for precise control and efficient operation of the SRM. Here's an overview of the classification:

4.4.1 Resolver

A resolver is an electromagnetic device that provides an analog output signal representing the rotor position. It's known for its accuracy and robustness in harsh environments, but it can be relatively bulky and expensive.

4.4.2 Encoder

Encoders are digital devices that provide discrete position information. They can be either incremental or absolute. Incremental encoders provide information about changes in position, while absolute encoders provide the exact position within a full rotation.

4.4.3 Hall Effect Sensors

Hall sensors detect the presence of a magnetic field. Placed on the stator or rotor, they provide information about the rotor's magnetic alignment relative to the stator poles. They're simple, cost-effective, and suitable for applications with moderate accuracy requirements.

4.4.4 Resolvers with Digital Output

These sensors combine the accuracy and robustness of traditional resolvers with the digital output of encoders. They provide a digital signal representing the rotor position, which simplifies integration with digital control systems.

4.4.5 Optical Encoders

Optical encoders use light-sensitive components to detect patterns on a rotating disc. They offer high accuracy and resolution, making them suitable for applications requiring precise control.

4.4.6 Inductive Sensors

These sensors detect the presence of metal using electromagnetic induction. They are reliable and durable, but their accuracy may be influenced by factors like temperature and material properties.

4.4.7 Rotary Variable Differential Transformer (RVDT)

RVDTs are electromechanical devices that provide analog output proportional to the rotor's angular displacement. They offer high accuracy but can be relatively complex and expensive.

4.4.8 Synchro Resolver

Similar to resolvers, synchro resolvers provide analog output corresponding to the rotor position. They're often used in systems where accurate position information needs to be transmitted between remote locations.

4.4.9 Magneto Strictive Sensors

These sensors exploit the magneto strictive effect to measure position. They offer high accuracy and resolution, making them suitable for demanding applications.

Using sensor feedback the close loop control has been design, one of the such close loop control is shown in Figure 4.3. The choice of rotor position

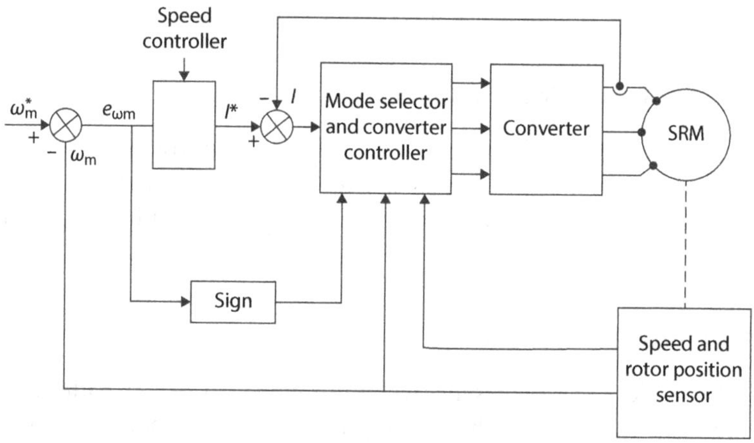

Figure 4.3 Schematic diagram of close loop control of SRM.

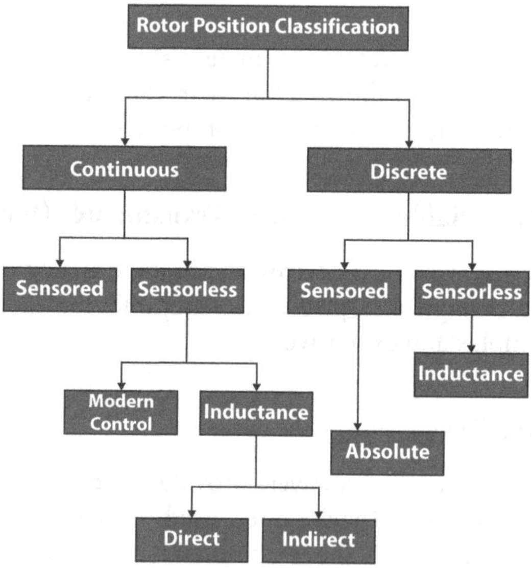

Figure 4.4 Classification sensing approach.

sensor depends on factors such as the required accuracy, environmental conditions, cost considerations, and the specific control strategy being employed in the SRM system. Further it can be classified as continues and discrete as shown in Figure 4.4.

4.5 Torque Equation of SRM

The torque, production in switched Reluctance motor control structures comes from the tendency of the rotor align with the excited stator poles. However, SRM has double salient poles and nonlinear magnetic characteristic. hence, the torque Ripple is more severe than those of other traditional motors. The torque ripple can be minimized through magnetic, circuit design and drive control. There is different method of control are available.

The controller operates by varying the turn on and turn off angles at every instant as per the required of the operating point. When the operation of the motor shifts from PWM mode to single pulse Mode, the turn on advanced to cater to the torque demand as the overlapping of the is reduced. When there is increase of load or increase of speed the then on angle is advanced and then off angle is retarded to valance the new torque demand. The emphasis is made to maintain operating show that to maintain the operating condition. Turn on and turn off angles vary to make the transition smooth between two, the two optional control modes. It is proved now that the developed controller able to control the SRM over entire speed and torque. schematic diagram of hysteresis current control is depicted in Figure 4.5.

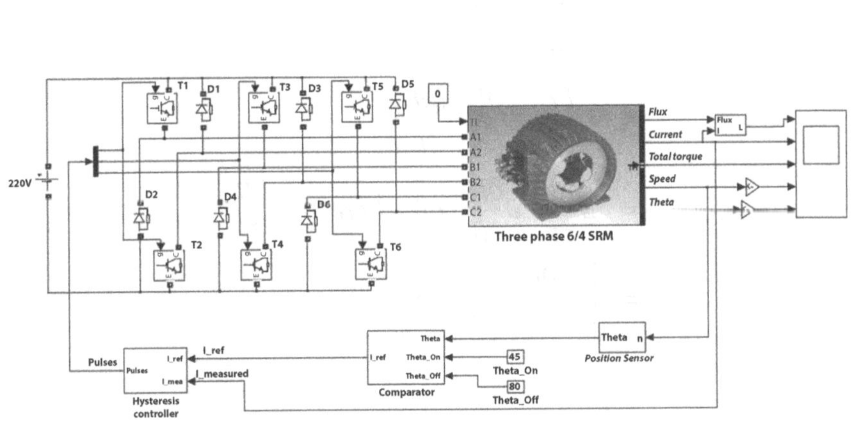

Figure 4.5 Hysteresis current control of SRM [2].

The controller operates by varying the turn on and turn off angles at every instant as per the required of the operating point. When the operation of the motor shifts from PWM mode to single pulse Mode, the turn on advanced to cater to the torque demand as the overlapping of the is reduced. When there is increase of load or increase of speed the then on angle is advanced and then off angle is retarded to valance the new torque demand. The emphasis is made to maintain operating show that to maintain the operating condition. Turn on and turn off angles vary to make the transition smooth between two, the two optional control modes. It is proved now that the developed controller able to control the SRM over entire speed and torque.

The torque is produced by variable reluctance principal. Flux linkage due to excitation of the winding is given by Eq. (4.1),

$$\lambda = Li \tag{4.1}$$

According to faradays law of electromagnetic inductance of emf(e) due to charge in flux linkages is given by Eq. (4.2),

$$e = (-)\frac{d\lambda}{dt} \tag{4.2}$$

Substituting λ from Eq. (4.1) in (4.2) gives Eq. (4.3)

$$\begin{aligned} e &= (-)\frac{d(Li)}{\partial t} \\ &= -L\frac{\partial i}{\partial L} - i\frac{\partial L}{\partial t} \\ &= -L\frac{\partial i}{\partial t} - i\frac{\partial L}{\partial \theta}\left(\frac{\partial \theta}{\partial t}\right) \\ &= -L\frac{\partial i}{\partial t} - i\frac{\partial L}{\partial \theta}\times\omega\left[where\ \omega = \frac{\partial \theta}{\partial t}\right] \\ &= -\left[L\frac{\partial i}{dt} + i\frac{\partial L}{\partial \theta}\cdot\omega\right] \end{aligned}$$

considering only magnitude θ, Eq. (4.4) is given as,

$$\theta = L\frac{\partial i}{\partial t} + i\frac{\partial L}{\partial \theta}\omega \tag{4.4}$$

Energy stored in magnetic field is given as Eq. (4.5),

$$W_e = \frac{1}{2}L\,i^2 \tag{4.5}$$

Power input to the motor calculated as per Eq. (4.6),

$$\begin{aligned} \mathrm{P} &= e_i \\ &= \left[L\frac{\partial i}{\partial t} + i(\omega)\frac{\partial L}{\partial \theta}\right]i \\ &= Li\frac{\partial i}{\partial t} + i^2\omega\frac{\partial L}{\partial \theta} \end{aligned} \tag{4.6}$$

Power due to variation in stored energy using Eq. (4.5) solved by $\frac{\partial we}{dt}$ and given as Eq. (4.7),

$$\begin{aligned} \frac{\partial we}{dt} &= \frac{dt}{dt}\left(\frac{1}{2}[i^2]\right) \\ &= \frac{1}{2}\left[2Li\cdot\frac{\partial i}{dt}\right] + \frac{1}{2}\left[i^2\frac{\partial L}{dt}\right] \\ &= Li\frac{\partial i}{\partial t} + \frac{1}{2}i^2\frac{\partial L}{dt} \\ &= Li\frac{\partial i}{dt} + \frac{i^2}{2}\frac{\partial L}{\partial 0}\left(\frac{\partial d}{dt}\right) \\ &= Li\frac{\partial i}{\partial t} + \frac{i^2}{2}(\omega)\frac{\partial L}{\partial \theta} \end{aligned} \tag{4.7}$$

subtracting (4.7) from (4.6), i.e. (4.6)-(4.7), Mechanical Power developed by

$$= P_m = \frac{1}{2}\left[i^2\omega \frac{\partial L}{J\theta} \right]$$

$$P_m = \frac{2\pi NT}{60} = \left[\frac{2\pi N}{60} \right] T = \omega T$$

$$P_m = \omega T \tag{4.8}$$

Where ω: angular velocity, T: Electromagnetic torque developed

$$\therefore T = \frac{P_m}{\omega} \tag{4.9}$$

substituting (4.9) in (4.8), torque is given by Eq. (4.10),

$$T = \frac{\frac{1}{2} i^2 \cdot \omega \frac{\partial L}{\partial \theta}}{\omega} \tag{4.10}$$

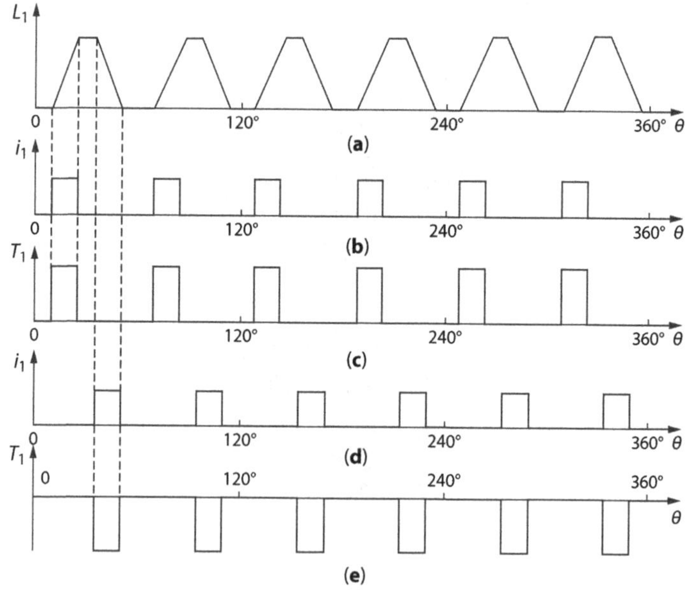

Figure 4.6 Inductance variation with angle.

Torque corresponds to motoring when $\frac{\partial L}{\partial \theta}$ is positive and torque corresponds to generating when $\frac{\partial L}{\partial \theta}$ is negative.

As T is proposal to i^2. Torque is indepordent of direction of current. Torque should be derived as the derivative with respect to rotor space position as Eq. (4.11).

$$\therefore T = \frac{1}{2}\left[i^2 \frac{\partial L}{\partial \theta} \right] \tag{4.11}$$

The inductance variation with rotor position is plotted in Figure 4.6.

4.6 Different SRM Converters

4.6.1 Two Power Semi-Conductor Switching Devices Per Phase and Two Diodes

Winding A, connected to a DC supply through specific components, activates devices based on the rotor position. When phase winding A is energized, devices T1 and T2 are activated according to the shaft position. To disconnect or de-energize phase winding A, the control circuit utilizes a shaft position sensor to deactivate devices T1 and T2. The energy stored in winding A sustains the current flow, directed through winding and components D1 and D2, back to the supply. The operational configuration, illustrated in Figure 4.7, demonstrates the circuitry for Mode-1 operation [3].

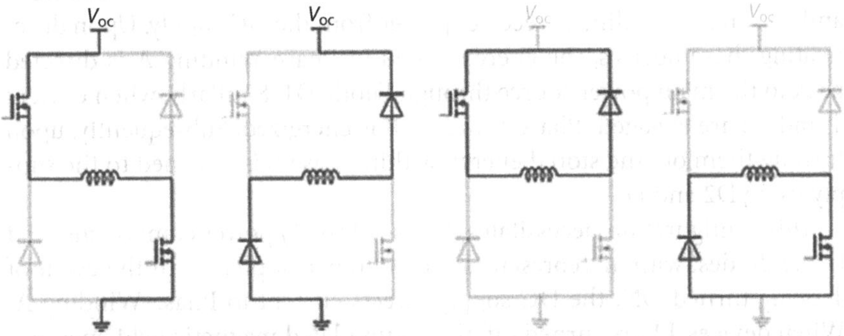

Figure 4.7 Operation mode-1 of two power semi-conductor switching devices.

Similarly phase B and C are also switched ON and OFF. From the supply in cyclic manner usually the upper devices T1, T3 and T5 are turned on and OFF from the signals obtained from the rotor position sensor.

The current in phase winding is the result of logical AND operation of the rotor position sensor and chopping frequency. As a result, it is possible to vary the effective phase current from a very low to very high value. For the following methods are available:

(1) The varying the duty cycle of chopper.
(2) By varying the conduction angle of the device.

4.6.2 Merits

- Control of each phase is completely independent of the other phase of the converter is able to free wheel during chopping period at low speeds which help to reduce the switching frequency and thus the switching losses of the convertor.
- The energy from the off going phase is feedback to the source which results in utilisation of energy is increase in efficiency.

4.6.3 Demerit

- High number of switching devices are used per phase, hence costly. Hence used for low voltage

4.7 (n+1) Double Switching Devices and (n+1) Diode

The depicted circuit employs a pair of switching devices and diodes, as illustrated in the above Figure 4.8. During the activation of SCRS switches T and T1, Phase Winding A receives power from the DC supply. Upon deactivating these devices, the energy stored in Phase Winding A is directed back to the main power source through Diode D1. Similarly, when devices T and T2 are engaged, Phase Winding B is energized. Subsequently, upon turning them off, the stored energy within Phase B is returned to the supply using D2 and D.

This configuration necessitates the use of (n+1) power components and (n+1) diodes, with 'n' representing the number of phases. In the event of T being turned ON, the DC supply directs current to Phase Winding A. When devices T1 are turned off, the accumulated magnetic field energy is channelled back to the supply via the winding and D1.

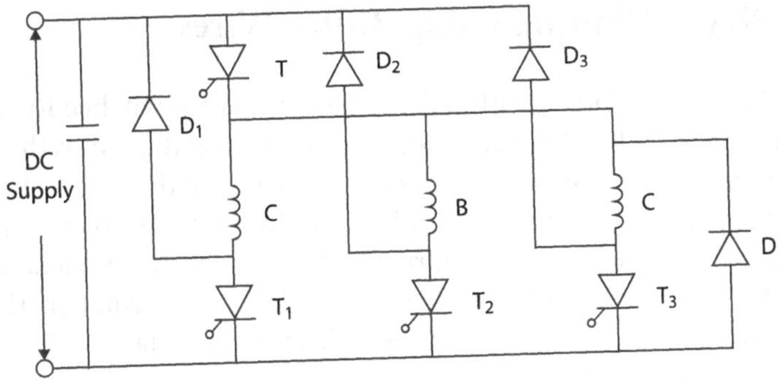

Figure 4.8 Circuit diagram of (n+1) double switching devices and (n+1) diode.

The three diodes operate in depending upon the signals obtained from the rotor position sensor and the chapping signals for PWM technique obtained from the controller [4].

4.7.1 Merits

- Converter uses low number of devices which reduces cost of the converter.
- The converter can free wheel during chopping. Thus, reduces switching frequency and losses.
- Voltage rating of all switching devices and diodes is Vdc which is relatively low.
- Energy for the off going phase is transfer back to source which results in use full utilization of the energy and also improves the efficiency.

4.7.2 Demerits

- Disability to magnetize a phase when off going phase is still de-magnctizing. This result in high torque ripple.
- At higher speed of the off going phase cannot be demagnetize fast enough because common switch keeps turning ON intermediately. Disabling force de-magnetization. The common switch conducts for all the phases. Thus, it has higher switching stress.

4.8 Phase Winding Using Bifilar Wires

Employing bifilar winding within the phase circuit of a Switched Reluctance Motor (SRM) entails a distinct winding configuration that holds the potential to augment the motor's operational efficiency and control attributes. The bifilar winding technique involves the intertwining of two parallel wires, as visually depicted in Figure 4.9. This approach can yield benefits such as heightened coil symmetry and a decrease in electromagnetic interference. Below is an elucidation of the utilization of bifilar winding within an SRM's phase circuit:

In a standard SRM phase winding, each coil is wound using a single wire. However, when employing bifilar winding, two wires are wound in parallel, maintaining consistent spacing and tension between them.

4.8.1 Merits

This technique offers several advantages:

1. Reduced Common-Mode Noise: Bifilar winding helps in cancelling out common-mode noise and interference that could affect the motor's operation. The two parallel wires generate equal and opposite currents when subjected to common-mode disturbances, minimizing their impact on the motor's performance.
2. Enhanced Symmetry: Bifilar winding enhances the symmetry of the coil arrangement. This symmetry can lead to better

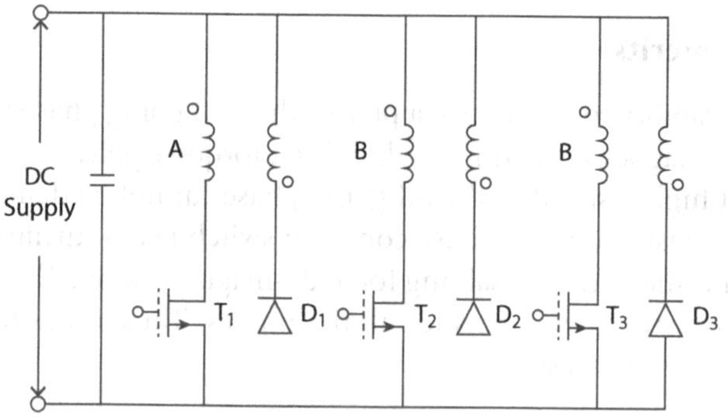

Figure 4.9 Circuit diagram of phase winding using bifilar wires.

control of the magnetic field within the motor, resulting in smoother operation and reduced torque ripple.
3. Improved Cooling: The parallel arrangement of wires allows for better heat dissipation. Heat generated during operation can be distributed more evenly across the winding, reducing the risk of localized overheating.
4. Reduced EMI: Electromagnetic interference (EMI) can be mitigated with bifilar winding due to the cancellation of induced voltages caused by external magnetic fields. This is particularly important in applications where EMI can adversely affect nearby electronic components.
5. Higher Current Capacity: Bifilar winding effectively doubles the cross-sectional area of the winding, allowing for increased current-carrying capacity. This can lead to higher torque output and power density.

4.8.2 Demerits

1. Bifilar winding entails a duplication of connection points.
2. It results in suboptimal copper utilization.
3. Free-wheeling does not occur in a seamless manner.
4. There's an increased occurrence of ripple.
5. The deficiency in coupling between windings leads to voltage spikes during the OFF state.
6. The chopper-less operation linked with the auxiliary winding becomes unacceptably elevated for numerous applications.

4.9 Split Link Circuit Used with Even Phase Number

4.9.1 Operation

Thc arrangement displayed in Figure 4.10 finds application in a range of highly efficient devices, spanning power capacities from 4 to 80 kW. The primary power supply undergoes division into two segments through the use of split capacitors. During the conduction phase, energy is fed to the phases via one half of the power supply. In the commutation interval, the phases demagnetize into the other half of the power supply.

When switch T is activated, phase winding 1 receives energy from capacitor C. Subsequently, when T4 is deactivated, phase winding 4 becomes energized.

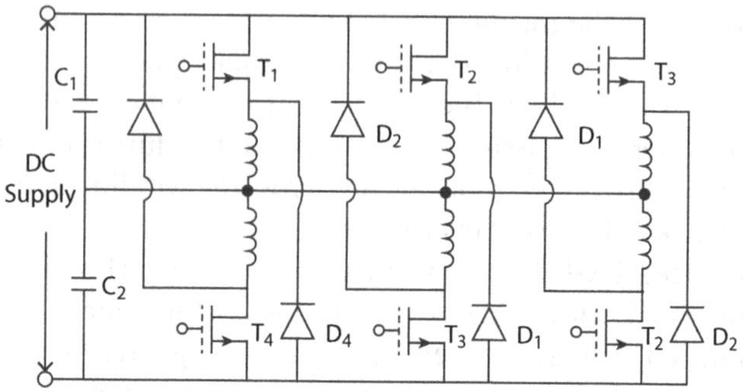

Figure 4.10 Circuit diagram of split link circuit used with even phase number.

Upon turning off T4, the energy stored in winding 4 is directed back to capacitor C through diode D. This sequence repeats analogously across the remaining windings as shown in Figure 4.11.

In the C dump circuit, the device count is reduced to n+1 additional devices to bleed the stored energy from the capacitor C back to supply via the steps down chopper circuit. The mean capacitor voltage is maintained well above the supply to permit rapid de-fluxing after commutation.

4.9.2 Merits

- A control failure in the energy recovery circuit would result in the rapid build-up charge on the capacitor and if protective measures were not taken, the entire converter would fail from over voltage.

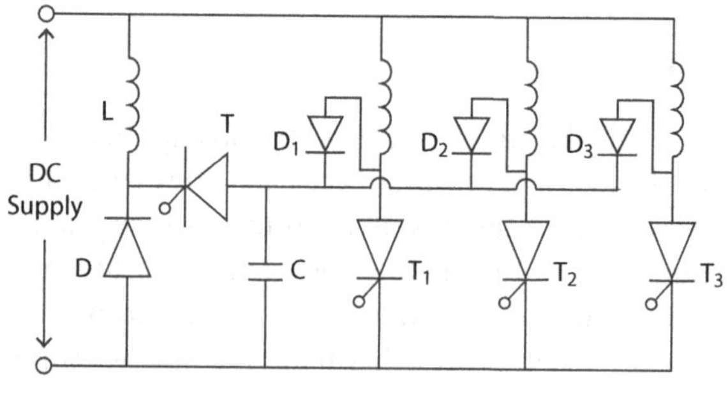

Figure 4.11 Switching circuit [4].

4.9.3 Demerit

- Introduce distortion on AC side.
- Switching circuit useful for starting current control.

4.10 Merits and Demerits of SRM

4.10.1 Merit

1. Exceptionally high torque-to-weight ratio, surpassing conventional alternatives.
2. Virtually eliminates the concern of temperature elevation.
3. Elegantly simplistic design due to the absence of permanent magnets.
4. Exemplary efficiency negates the necessity for external cooling.
5. Completely devoid of commutators and brushes, thereby ensuring low maintenance requirements.
6. Rotor's lack of windings, slip rings, and brushes contributes to a maintenance-free operational life.
7. Remarkably resilient against shoot-through faults, bolstering operational reliability.
8. Torque production remains independent of current polarity, enhancing versatility.
9. Facilitates seamless transitions between forward, reverse, acceleration, deceleration, and braking operations.

4.10.2 Demerits

1. The stator phase winding must accommodate the magnetizing current to establish the necessary airgap flux.
2. At elevated speeds, the generated torque exhibits undesirable fluctuations, introducing ripples in performance.
3. Optimal motor design necessitates a tight airgap along the aligned axis, while a significantly larger airgap is required at the inter-pole axis.
4. A lack of standardized practices adds complexity to design and implementation.

5. The motor's dimensions are comparable to those of variable-speed induction motor drives, possibly posing space constraints and installation challenges.

4.11 Applications

1. Elevating Vertical Mobility: SRMs are integral to high-speed elevators, enhancing vertical transportation efficiency. They also optimize washing machines and vacuums, delivering precision and energy savings to daily tasks.
2. Pioneering Automotive Progress: SRMs are a cornerstone of the automotive future, especially for electric vehicles (EVs). Their exceptional torque, efficiency, and adaptability are propelling the promising trajectory of EV technology.
3. Empowering Robotics: SRMs enable precise control in robotic systems, ensuring accurate movement and versatile task execution across industries.
4. Precision in Food Processing: In the realm of food processing machinery, SRMs offer precision and reliability, ensuring consistent and efficient production.
5. Aviation Advancements: SRMs contribute to aviation by enhancing aircraft systems with their controllable torque and adaptable performance.
6. Effective Traction: In traction applications, SRMs play a vital role, providing efficient power delivery for various modes of transportation.
7. Propelling Maritime Innovation: SRMs find utility in ship propulsion systems, showcasing their potential in maritime operations.
8. Leveraging Technological Synergy The convergence of AI, finite element analysis software, advanced computers, and simulation tools like MATLAB Simulink facilitates seamless SRM design. This technological synergy simplifies the creation of optimized designs, propelling the field forward.

4.12 Future Scope

The potential of this motor is vast, spanning from applications at speeds as high as 1 lakh to 2 lakh RPM, to scaling up for megawatt-level power requirements. Its versatility makes it an excellent fit for electric vehicles and traction systems, complementing its compatibility with conventional uses. Moreover, the motor holds immense promise for innovative research, facilitated by contemporary facilities.

A significant development is the emergence of HR-SRM, with TURNTIDE technology USA leading the charge. This technology has brought forth motors with ratings up to 200 KW, showcasing an impressive 96% efficiency. This not only exemplifies the motor's capabilities but also underlines its adaptability for demanding industrial needs.

In essence, the motor's future prospects are characterized by its adaptability across various RPM ranges, power scales, and applications. Electric mobility, traction systems, and conventional uses are only the beginning; the motor's untapped potential invites pioneering research and innovation. With HR-SRM technology setting new efficiency standards, it's evident that the motor is poised to make significant contributions across diverse domains.

Conclusion

In conclusion, the chapter delved into a comprehensive analysis of Switched Reluctance Motors (SRMs). It covered various aspects including constructional features, operational principles, torque-speed profiles, drive circuitry, and provided real-world examples of SRMs in action. The significance of grasping these concepts cannot be overstated, as SRMs have gained prominence as efficient and adaptable motors in various industrial applications. A solid understanding of their design, behavior, and control strategies is essential for harnessing the full potential of SRMs and integrating them effectively into modern engineering practices.

References

[1] X. D. Xue, K. W. E. Cheng, S. L. Ho. "Improvement of Power Factor in Switched Reluctance Motor Drives Through Optimizing the Switching Angles," Electric Power Components and Systems, 2004.
[2] H. Goto, "A multipolar SR motor and its application in EV," Journal of Magnetism and Magnetic Materials, vol. 2005, 2005.
[3] S. Mir, "Classification of SRM Converter Topologies for Automotive Applications," SAE International, 2000.
[4] Janardanan, E. (2014). Special Electrical Machines. India: PHI Learning.

5

Stepper Motors

Ankur P. Desai[1] and Jignesh P. Desai[2]*

[1]GIDC Degree Engineering College, Navsari, India
[2]U. V. Patel College of Engineering, Ganpat University, Mehsana, Gujarat, India

Abstract
The stepper motor chapter provides a brief yet comprehensive overview of the fundamental principles and applications of stepper motors. Exploring their magnetic field-driven step-wise movement, the chapter highlights their crucial roles in robotics, automation, and precision applications. It touches upon the distinctive features of stepper motors, including open-loop control and precise positional accuracy, while also discussing recent advancements in driver electronics and control algorithms. Overall, it serves as a concise guide for understanding the core principles and practical applications of stepper motor technology.

Keywords: Step angle, stepper motor, coil, drive, speed control

5.1 Introduction

Because of their precise control over angular position and speed, stepper motors are widely used in a variety of industries and applications. In this chapter, we will look at the constructional features, operating principles, torque-speed characteristics, drive circuits, and examples of stepper motors.

5.2 Constructional Features

Stepper motors are made up of a number of key components that work together to produce accurate and controlled rotational motion. The rotor,

**Corresponding author*: jpd.fetr@gmail.com

Jigneshkumar P. Desai (ed.) Special Electrical Machinery, (99–106) © 2024 Scrivener Publishing LLC

Figure 5.1 View of a stepper motor.

stator, pole pieces, and bearings are the main components of a stepper motor.

The rotor is the stepper motor's moving component that generates the output motion. A permanent magnet or a soft iron core with teeth known as rotor poles is typically used. The design and construction of the rotor are critical in determining the motor's torque and step resolution. The cross section of stepper motor is shown in Figure 5.1.

The stator is the stepper motor's stationary component that surrounds the rotor. When energised, it is made up of multiple coils or windings that are placed in specific positions to create a magnetic field. The arrangement of the stator and the number of poles have a significant impact on the motor's step resolution and overall performance.

A stepper motor's pole pieces are strategically placed to interact with the rotor poles and create magnetic fields. The interaction of the magnetic fields of the rotor and stator produces the torque required for rotor movement.

Stepper motors also require bearings, which are an important component. They support the rotor shaft and reduce friction, allowing for smooth rotation. High-quality bearings ensure that the motor operates with the least amount of mechanical resistance, improving overall performance and lifespan.

5.3 Principle of Operation

The operation of a stepper motor is based on the concept of electromagnetic fields and the interaction of the rotor and stator. Stepper motors use electromagnetic induction and the concept of step resolution to operate.

Electromagnetic induction: A magnetic field is created when an electric current flows through a coil. The interaction of the magnetic fields of the rotor and stator produces torque, which causes the rotor to move. The magnetic fields are manipulated to produce precise stepwise motion by controlling the energization and de-energization of the stator coils.

Step Resolution: Stepper motors are capable of rotating in fixed angular increments known as steps. The resolution of the motor is determined by the number of steps per revolution. Step resolutions of 200, 400, or 800 steps per revolution are common. Higher step resolutions provide finer control over the position of the motor but may necessitate more complex drive circuits.

Stepper motors rely on sequencing the energization of the stator coils to achieve controlled motion. The magnetic fields interact with the rotor's poles, causing it to move step by step, by energising the coils in a specific sequence. The resolution and smoothness of different sequencing methods, such as full-step, half-step, and microstepping, vary.

To visually explain the principle of operation, detailed illustrations and diagrams can be provided, demonstrating the interaction between the rotor, stator, and magnetic fields during each step. Figure 5.2 shows the working of stepper motor.

In this Table 5.1, the "Step" column represents the step number, and the "Coil A," "Coil B," "Coil C," and "Coil D" columns indicate the energization state of the respective coils. The "Step Angle" column indicates the corresponding angle in degrees for each step. This table provides a mapping between the step number and the coil energization pattern in a stepper motor, along with the resulting step angle for each step [1].

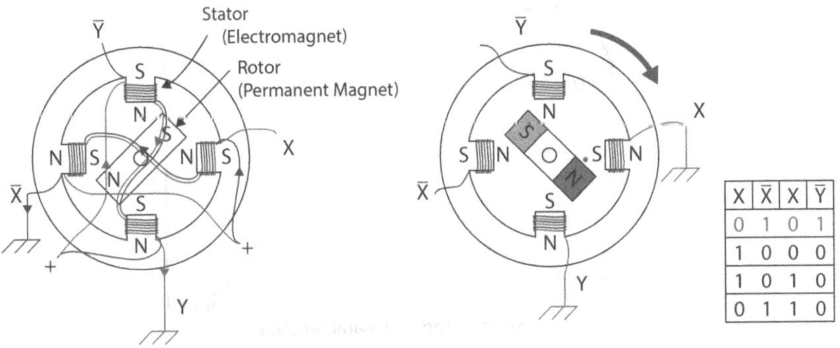

Figure 5.2 Stepper motor working principle.

Table 5.1 Operation of stepper motor.

Step	Coil A	Coil B	Coil C	Coil D	Step angle
0	1	0	1	0	0°
1	0	0	1	0	90°
2	1	0	0	1	180°
3	0	1	0	1	270°

5.4 Torque-Speed Characteristics

A stepper motor's torque-speed characteristics are determined by several factors, including the motor's construction and the drive circuit used. Understanding these characteristics is critical when choosing the right stepper motor for a given application [2].

The holding torque is the maximum torque that a stepper motor can generate when it is stationary.

It determines the motor's ability to maintain a position in the absence of external force. Higher holding torque values are preferred for applications that require precise positioning. The torque verses speed characteristic is depicted in Figure 5.3.

Pull-in Torque: The pull-in torque is the minimum torque required to start a rotor from a stop. It determines the ability of the motor to start and overcome static friction. The pull-in torque must be strong enough to overcome any resistive forces and start the motion smoothly.

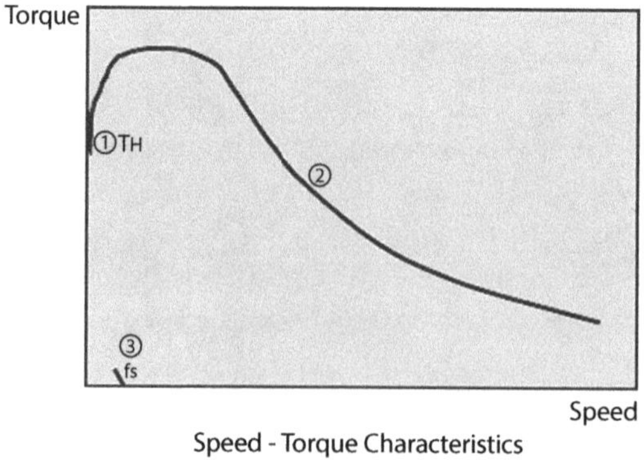

Figure 5.3 Torque-speed characteristics of a stepper motor.

Pull-in Torque: This is the minimum torque required to start a rotor from a standstill. It determines the motor's ability to start and overcome static friction. The pull-in torque must be sufficient to overcome any resistive forces and allow the motion to begin smoothly.

To demonstrate the relationship between torque and speed, detailed torque-speed curves and graphical representations can be included, emphasising the motor's capabilities under various operating conditions.

5.5 Drive Circuits

To illustrate the relationship between torque and speed, detailed torque-speed curves and graphical representations can be included, emphasising the motor's capabilities at various operating conditions. The drive circuit is shown in Figure 5.4 and Figure 5.5.

Unipolar Drive: Unipolar drive circuits control the flow of current through the stator windings using a power supply and switching elements. They are relatively easy to implement but may be inefficient. Unipolar drives are appropriate for applications with low torque requirements and a preference for simplicity.

Bipolar Drive: Bipolar drive circuits control the direction and magnitude of current flowing through the stator windings using H-bridge configurations or specialised driver ICs. They provide more torque and efficiency, but are more complicated. Bipolar drives are preferred in applications that require more torque and precision.

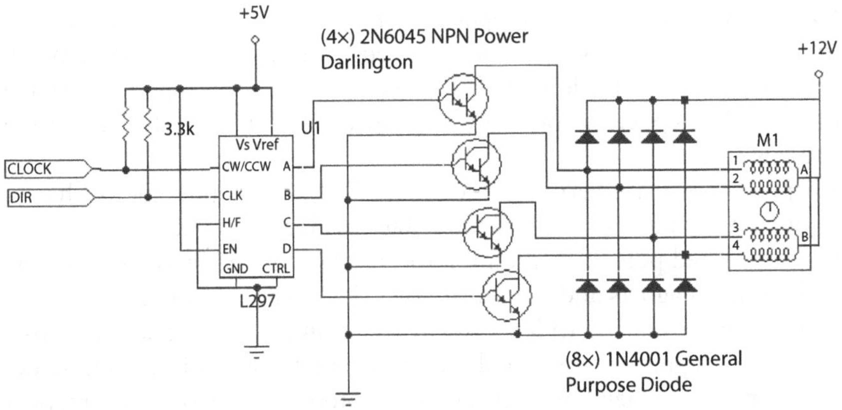

Figure 5.4 Unipolar drive circuit.

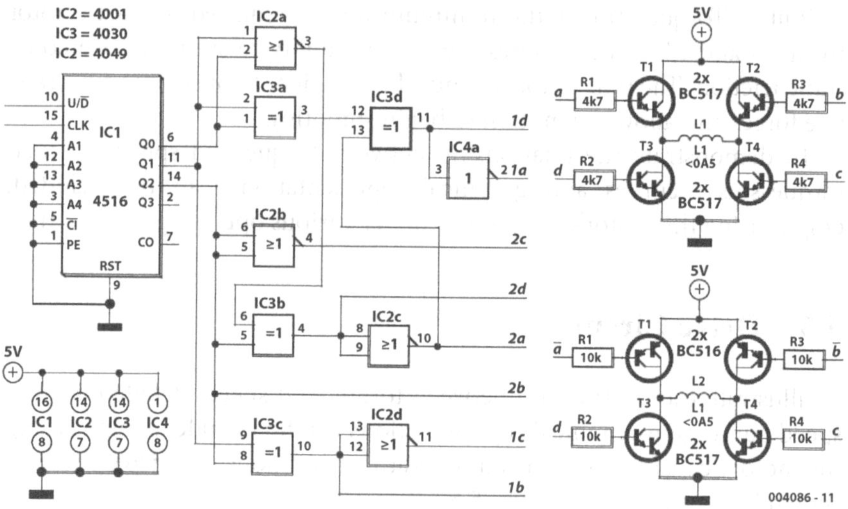

Figure 5.5 Bipolar drive circuit.

To provide a comprehensive understanding of the drive systems used in conjunction with stepper motors, detailed circuit diagrams and explanations of both unipolar and bipolar drive circuits can be included.

5.6 Types of Steppers Motors

Stepper motors are classified into several types based on their design and operation. Some examples of commonly used types are:

Variable Reluctance (VR) Stepper Motors: These motors are made of soft iron teeth on the rotor and concentrated windings on the stator. When energised, the rotor aligns with the stator poles, resulting in rotational motion. VR stepper motors are well-known for their high step resolution and low cost.

Stepper Motors with Permanent Magnets (PM): PM stepper motors have a rotor with permanent magnets and a stator with either concentrated windings or teethed poles. The required torque is generated by the interaction of the rotor magnets and the stator poles. PM stepper motors have a high holding torque and are widely used in precision positioning applications.

Hybrid stepper motors combine the characteristics of variable reluctance and permanent magnet stepper motors. They have a permanent magnet rotor and a stator with teethed poles. Hybrid stepper motors offer a good balance of step resolution, torque, and cost, making them suitable for a variety of applications.

5.7 Example: Calculation of Steps per Revolution

1. Assume we have a hybrid stepper motor with 200 steps per revolution (s) and a linear motion lead screw mechanism. We want to know how many steps are required to move the load a certain distance.

Solution:
The formula to determine how many steps (N) linear motion needs is as follows:

$$N = (D * s) / (P * M)$$

D stands for the intended linear travel distance, P for lead screw pitch, and M for microstepping resolution.

Consider the following parameters, which represent 1/8th microstepping resolution: D = 100 mm, P = 5 mm, and M = 8.

N is equal to $(100\,mm * 200\,steps/rev) / (5\,mm/rev * 8)$, which equals 5000 steps.

With the included lead screw and stepper motor, 5000 steps are needed to accomplish a linear displacement of 100 mm.

5.8 Applications

Numerous devices, such as robots, CNC machines, 3D printers, and automation systems, employ stepper motors. Several instances of their adaptation are shown below:

1. Robotics: To accurately regulate joint motions in robotic systems, stepper motors are frequently utilised. They make robots more accurate, repeatable, and flexible, making them appropriate for home robotics projects, medical devices, and industrial automation.
2. Computer numerical control (CNC) machines: In CNC machines, the movement of the machine's axes is managed by stepper motors. Stepper motors provide complicated

machining processes like a milling machine, lathe, or 3D printer the precision and control they need.

3. 3D printers: To regulate the movement of the print head and the location of the print bed, stepper motors are extensively utilised in the additive manufacturing process. Stepper motors enable accurate layer-by-layer deposition, resulting in the creation of 3D printed items of excellent quality.
4. Automation Systems: Stepper motors are frequently used in automated systems such robotic assembly lines, packing equipment, and conveyor systems. They are perfect for applications demanding repeated motion and accurate operation because they may offer exact positioning and control.

5.9 Conclusion

The chapter involved an exploration of the structural attributes, operational concepts, torque-speed behaviors, drive circuits, and instances of stepper motors. Understanding these principles holds significant importance.

References

1. Janardanan, E. (2014). Special Electrical Machines. India: PHI Learning.
2. Sahdev, S. K. (2017). Electrical Machines. United Kingdom: Cambridge University Press.

6

Universal Motor

Jigneshkumar P. Desai

U. V. Patel College of Engineering, Ganpat University, Mehsana, Gujarat, India

Abstract

The universal motor chapter succinctly covers the principles and applications of universal motors. These versatile motors operate on both AC and DC power, offering high-speed performance and a compact design. Widely used in appliances and power tools for their cost-effectiveness and high power-to-weight ratios, universal motors are discussed with a focus on their advantages and recent developments. The chapter concludes by emphasizing their enduring relevance across various industries.

Keywords: Construction, operation, pole, universal

6.1 Introduction

The universal motor is a multifunctional and often employed electrical device that can run on both AC and DC power sources. As a result of its small size, high power density, and speed control skills, it finds use in a variety of sectors. We will look at the concentrated-pole, non-compensated type, speed control techniques, and applications of the universal motor in this chapter.

6.2 Construction

A series DC motor's structure is similar to that of a universal motor. It is made up of a rotor with armature windings and a stator with field windings. The field windings and armature windings are commonly linked in

Email: jpd.fetr@gmail.com

Jigneshkumar P. Desai (ed.) Special Electrical Machinery, (107–112) © 2024 Scrivener Publishing LLC

Figure 6.1 Cross-section of a Universal motor.

series, enabling the motor to run on both AC and DC power sources. The commutator segments and brushes make it easier for current to flow to the rotor windings, and the armature core is laminated to prevent eddy current losses. The cross section of Universal motor is shown in Figure 6.1.

6.3 Principle of Operation

The interaction between the magnetic field generated by the field windings and the current flowing through the armature windings is the basis of the universal motor's working principle. An armature experiences a mechanical force when an electric current flows through the windings, creating a magnetic field in the process. The rotor rotates as a result of this force, also known as the Lorentz force. Figure 6.2 shows the operation and connection of universal motor.

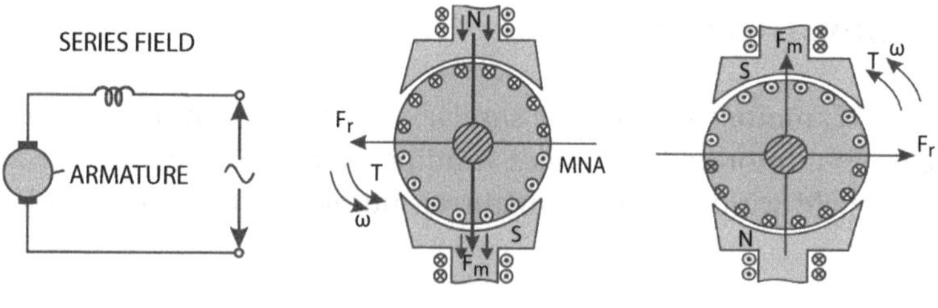

Figure 6.2 Principle of operation of a Universal motor [1].

6.4 Concentrated-Pole-Non-Compensated Type

The interaction between the magnetic field generated by the field windings and the current flowing through the armature windings is the fundamental working principle of a universal motor. A magnetic field created by an electric current flowing through the windings exerts a mechanical strain on the armature. The rotor rotates due to this force, sometimes referred to as the Lorentz force.

6.5 Speed Control Methods

For a universal motor to be used in a variety of sectors, speed control is essential. Torque verses speed characteristic is depicted in Figure 6.3. A universal motor's speed can be managed using a variety of techniques, such as:

1. Voltage Control: The motor's speed may be changed by adjusting the input voltage. Lower voltages slow down the pace, whereas higher voltages increase it. Although this approach is straightforward and economical, it might reduce torque at slower speeds.
2. Armature Resistance Control: The speed of the motor may be adjusted by connecting external resistance in series with the armature windings. While reducing the resistance causes the speed to rise, increasing the resistance causes the speed to decrease. Although a greater range of speed control is possible with this approach, efficiency is decreased.

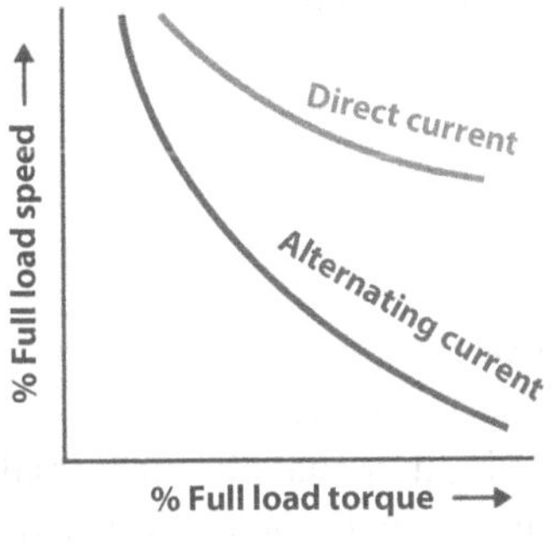

Figure 6.3 Characteristic of Universal Motor.

3. Electronic Speed Control: The accurate and effective speed control of universal motors is made possible by the employment of electronic speed control techniques like pulse-width modulation (PWM) or chopper control. These techniques change the effective voltage delivered to the motor while keeping the average voltage constant and modifying the waveform's duty cycle.

6.6 Applications

The versatility and small size of the universal motor allow it to be used in a variety of sectors. The following are some typical uses for the universal motor:

1. Home Appliances: Electric fans, blenders, mixers, and vacuum cleaners are just a few examples of the home appliances that frequently employ universal motors. They are perfect for various applications thanks to their small size, high power density, and speed control capabilities.
2. Power Tools: Universal motors are commonly found in power tools like drills, grinders, and routers. Their ability to operate on both AC and DC power sources, along with their high power output, allows for efficient and versatile tool operation.
3. Electric Vehicles: Due to its small size, high power density, and affordability, universal motors are utilised in tiny electric vehicles like electric bikes and scooters.
4. Industrial Machinery: Various industrial machinery, such as sewing machines, conveyor systems, and tiny machinery requiring variable speed control, find uses for universal motors [2].

Conclusion

The chapter covered the examination of concentrated-pole, non-compensated type, speed control methods, applications, and the assembly of the universal motor. It was highlighted that the universal motor, known for its adaptability and frequent employment across various industries, can operate using both AC and DC power sources. Its compact size and capacity for speed regulation further contribute to its popularity. Understanding the design, operation, and speed control approaches of the universal motor is essential for its effective utilization across a range of applications.

References

1. Sahdev, S. K. (2017). Electrical Machines. United Kingdom: Cambridge University Press.
2. Burman, S. P. (2013). Special Electrical Machines. India: S.K. Kataria & Sons.

7

Torque Pulsation Analysis in PMSM

Kannan Selvam[1]*, Subhanarayan Sahoo[1,2] and Kaumil Shah[3]

[1]*Gujarat Technological University, Ahmedabad, India*

[2]*Department of EE and Energy Innovation Center, Ganpat University, Mehsana, India*

[3]*Department of EE, U. V. Patel College of Engineering, Ganpat University, Mehsana, India*

Abstract

The Permanent Magnet Synchronous Motor (PMSM) has replaced the usage of Induction Motor (IM) or Brushless DC Motor (BLDCM) in many applications due to its higher performance characteristics such as efficiency, torque-to-current ratio, torque-to-volume ratio and power factor. But industrial and transportation applications require ripple free torque whereas PMSM exhibits adverse effect on the torque pulsation so that it's to highly be addressed. This paper presents the comprehensive review of the possible causes of torque pulsation of PMSM drives. The major design parameters which influence the cogging torque, switching methodologies of power converters and other causes such as braking and switching failures are analysed for the model of 1kW motor and simulation results are presented here.

Keywords: PMSM, cogging torque, flux remanence, CPMR, skewing, FOC, DTC, commutation delay

7.1 Introduction

The PMSM has proven the best performance compared to IM and BLDCM if torque, speed, efficiency and losses are considered [1]. Also PMSM have low noise and comparatively low torque pulsation which leads to ease of control. There are few major advantages over IM [2] which make the manufactures to choose the PMSM as mentioned below:

**Corresponding author*: engineerskannan@gmail.com

Jigneshkumar P. Desai (ed.) Special Electrical Machinery, (113–142) © 2024 Scrivener Publishing LLC

- The inertia of PMSM is low and torque response is high due to the absence of cage rotor and insignificant rotor losses.
- The IM requires higher rated power converter than PMSM when both have same rating of the output.
- The size of PMSM is smaller than IM for the same rating and it makes more compact in space-concerned applications.
- Also torque to volume ratio is higher than IM.

These advantages made the PMSM a best choice for rigid and promising performance in many applications such as aerospace actuation [3], Electric vehicles (EV's) [4], Industrial cranes [5], etc.

Especially for EV applications, PMSM is mostly considered than IM and Switched Reluctance Motor (SRM) [6]-[7], due to high performance characteristics of PMSM as mentioned in [8], [9]. A PMSM is similar to a 3-ϕ synchronous motor, but the excitation is provided by a Permanent Magnet (PM) in place of electric winding in rotor [9]. Especially in automobile applications, if electric traction is appointed then the motor must be capable of withstanding continuous start or stop and sudden acceleration or deceleration rates. The use of PMSM fulfils these requirements and also rigid enough to provide high controllability, stable and appropriate transient performance. But smooth control over torque pulsation is quite difficult and it must be addressed.

In general the torque pulsation is often termed as variation in the output torque with respect to the drives average torque, but in plain words it can be quoted as ripples over smooth torque output. The torque pulsation is usually undesirable because of vibration on drives and noise produced on the drive as well as it leads to reduction of machine lifetime [10]. This paper deals with all the major causes of torque pulsation occur in the PMSM and it helps to see the multi-angular view of torque pulsation as well as to design such applications.

Usually the construction of PMSM and control strategies used in drives are the two major parameters which contribute the torque pulsation most [11], [12]. The detailed classification of torque pulsation is presented in Figure 7.1. In manufacturing point of view, the variance of magnetic field and improper slot fixture affect the cogging torque and so does the torque pulsation. The vibration caused by torque pulsation is nullified by placing the high power magnet interlacing and changing the magnetic orientation at pole transition [13] as it reduce the amplitude of driving frequency in practice.

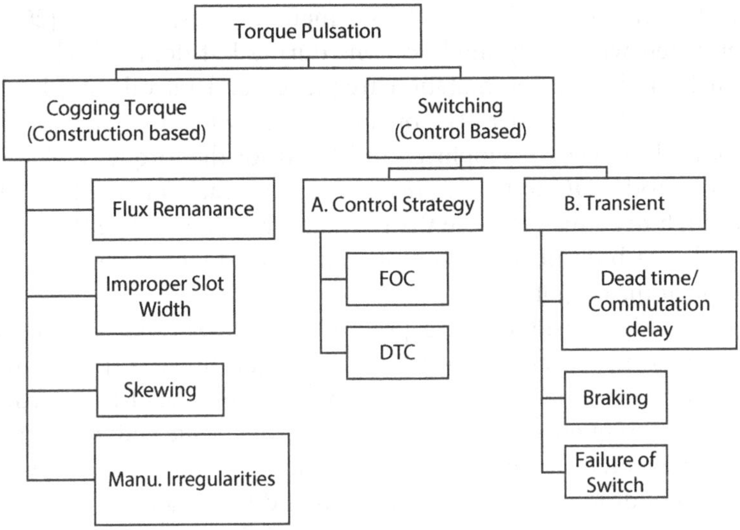

Figure 7.1 Classification of causes of torque pulsation.

When Permanent Magnets (PM) are exposed to higher temperature environments, the flux remanance gets altered and it directly impact on PM's flux linkage [14], followed by the variation in the cogging torque so does the torque pulsation. Even optimization of slot and pole number combination [15] is considered to reduce the cogging torque. While selection of fractional-slot winding explicit lower cogging torque, the average torque output has been reduced.

Furthermore, the improper slot width plays critical role in influencing the cogging torque [16]. The step-skewing of magnets [17] or slots of stator are also often followed for reducing the cogging torque, but selection of skewing angle has notable significance on torque pulsation. Also some manufacturing irregularities during mass production [16] such as dynamic eccentricities of stator and rotor misalignments may also lead to further variation in cogging torque [18].

In control strategic point of view, the switching techniques of power converter employed and transient conditions usually play the major role in torque pulsation. The Field-Oriented Control (FOC) for PMSM is introduced in [19] along with active filter topology to reduce the torque pulsation and harmonic noises. By detecting the harmonics present in the stator voltage, the active filter maintains the sinusoidal voltage in stator with the help of hysteresis controller in this approach.

The Direct Torque Control (DTC) method is explored in [20], and implemented with many modifications during last decade [21], [22]. In general by reducing commutation frequency and introducing low-pass filter-based voltage model [21], or modifying switching table, duty cycle control [23] the DTC is employed in PMSM for the torque ripple reduction. Even both FOC and DTC are compete with each other [24], the DTC proves high dynamic response than FOC as mentioned in [24], [25], [26] and [27]. But both of these control strategies exhibits significant torque pulsation at the output.

Though FOC produces less pulsation than DTC [24], controlling of settling time at sudden acceleration is questionable [27]. Further, the commutation delay of power converters and dead-time provided in switching sequences influence in a part of torque pulsation since it induces unwanted current harmonics as shown in [28] and [29]. Also the transient conditions such as acceleration and braking may have notable impact on the torque pulsation, especially in EV's [30]. EV's generally use the regenerative braking in order to save the energy, so during this flux reversal, a time delay is created which causes the torque pulsation due to the commutation interval [31].

Similarly, the failure of power electronic switches and diodes used in power converters may also affect the dynamic performance of the drive and sudden changes in output torque [32]. These are the most possible causes of torque pulsation in PMSM and to be analysed while constructing drives with optimized output and high performance. All issues have been discussed and presented along with supporting simulation works here.

7.2 Construction Based Torque Pulsation

A. Cogging Torque

The cogging torque is generally known as the summation of the interactions of each edge of the PM present in the rotor with the slot openings; and the analysis of these interactions may lead us to understand it well [16]. Since the cogging torque has fundamental frequency and zero sub-hormonics, it can be expressed [33] as follows using fourier sum:

$$T_C=\sum_{m=1}^{\infty} T_m \sin(mf_C\theta+\varphi_m) \tag{7.1}$$

where, T_m is peak value of cogging torque, Nm

m is order of cogging harmonics

f_c is cogging cycles per mechanical revolution of fundamental frequency

φ_m is determining the angular position with respect to the rotor reference

θ is determining the rotor displacement with respect to the stator reference

If we consider independent component for the 's' number of slots, then Eqn. (7.1) become,

$$T_C = \sum_{m=1}^{\infty} T_m \cos(ms\theta) \tag{7.2}$$

Also the components of cogging torque can be mentioned as the Cycles Per Mechanical Revolution (CPMR) [20], which exactly depends on the number of slots and poles. Considering Figure 7.2, if 's' number of slots and 'p' number of poles are present then the CPMR is computed [33] as,

$$\text{CPMR}\ (f_c) = \text{LCM}\ (s, p) \tag{7.3}$$

where LCM is least common multiple.

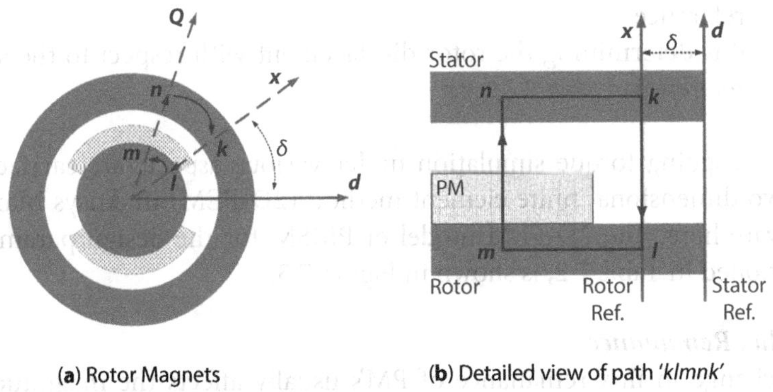

Figure 7.2 Basic structure of 2-pole surface mount PMSM.

Table 7.1 Typical values of CPMR [33].

s	p	f_c = LCM (s,p)
8	12	24
18	12	36
27	6	54
36	8	72

The typical values of CPMR for different configurations are mentioned in Table 7.1. In general when cogging cycles are higher, it leads to reduction of magnitude of cogging torque and mostly preferred. Further by calculating the co-energy around PM's edges to the nearest tooth and slot opening as shown in Figure 7.2, the cogging torque can be derived [33] and mentioned [34] as,

$$T_C(\theta) = -\frac{\partial W_f(\theta)}{\partial \theta} = -\frac{1}{2}\left(\frac{\partial \rho(\varphi,\theta)}{\partial \theta} \bullet F^2(\varphi,\theta) + 2F(\varphi,\theta) \bullet \frac{\partial F_o}{\partial \theta}\rho(\varphi,\theta)\right) \quad (7.4)$$

where, W_f is the magnetic field energy

$\rho(\varphi,\theta)$ is the permeance function

$F(\varphi,\theta)$ is the effective MMF in the magnetic system

F_0 is the MMF drops on the air gap

φ is determining the angular position with respect to the rotor reference

θ is determining the rotor displacement with respect to the stator reference

The cogging torque simulation under various aspects are carried out by two dimensional finite element method (2D-FEM) in Ansys Maxwell software here. The 2D-FEM model of PMSM for the design parameters mentioned in Table 7.2, is shown in Figure 7.3.

(i) Flux Remanance

The change in flux remanance of PM's usually affects the magnitude of cogging torque. The flux remanance gets altered when the PM's experience

Table 7.2 Design parameters.

Parameter	Value	
Rated power	1 kw	
Rated speed	1500 rpm	
Pole pairs	2	
No. of Slots	24	
Conductors per slot	52	
	Stator	***Rotor***
Outer diameter	120 mm	75 mm
Inner diameter	78 mm	25 mm
Length	60 mm	60 mm

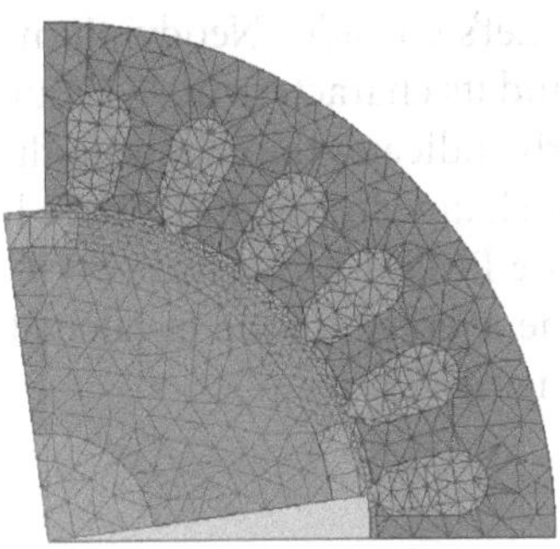

(a) ***2D FEM Model***

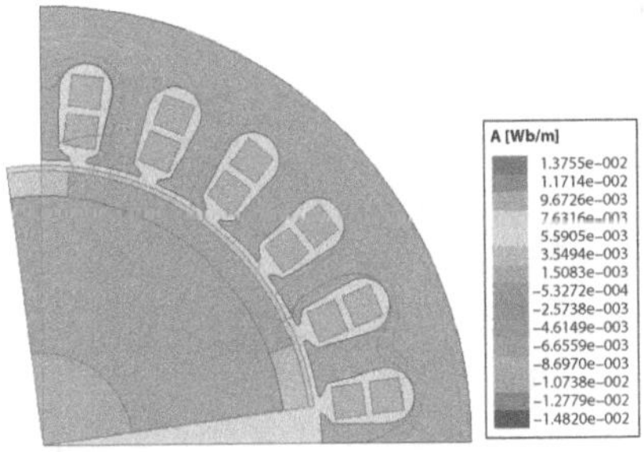

(b) ***Distribution of magnetic line***

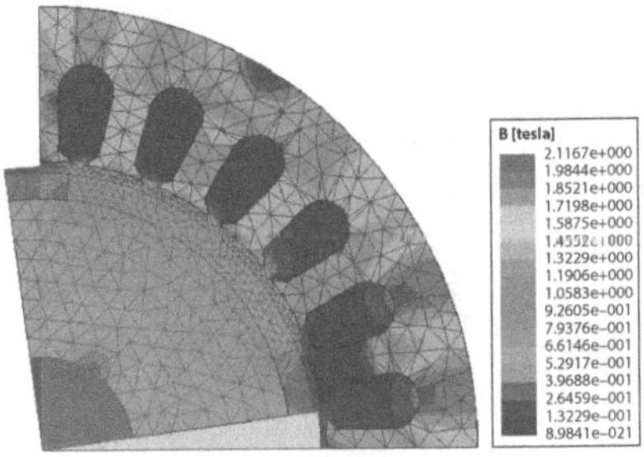

(c) ***Distribution of Flux density***

Figure 7.3 The 2-D FEM simulation of PMSM.

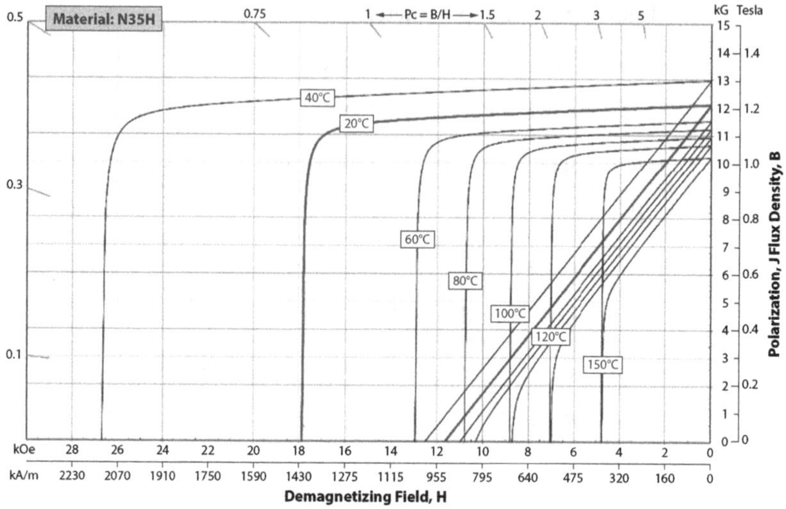

Figure 7.4 Characteristic B-H curve of N35H [35].

the higher temperature. Let's consider Neodymium PM (N35H) which is mostly used in PMSM, and its characteristic B-H curve [35] is given below in Figure 7.4, and it clearly indicates the change in flux density for different temperature, so does the change in cogging torque [36]. Usually the operational temperature of the PMSM varies from 90° C to 120° C in most of the applications [14]. The average torque is calculated in Ansys Maxwell software and shown in Figure 7.5.

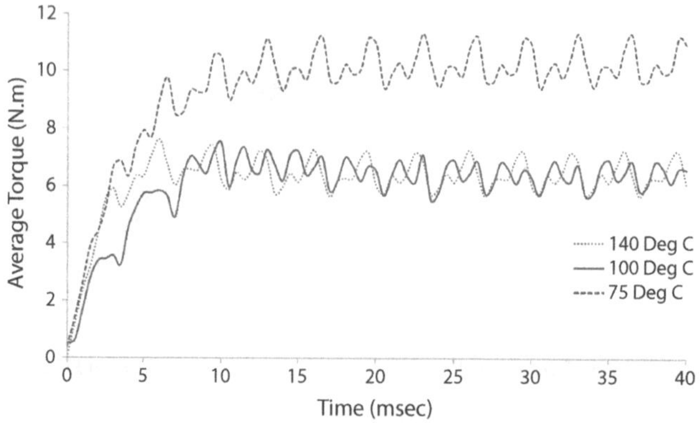

Figure 7.5 Analysis of output torque variation in PMSM with N35H magnets.

(Conversion 1kA/m=12.56 Oe, 1kOe=79.577 kA/m, 1T=10kGauss and 1 Gauss=0.1mT).

(ii) Improper Slot Width

The interaction of the magnets in rotor and slot openings from the stator usually determine the cogging torque. In order to achive optimized torque output, the slot openings are kept equal to width of the tooth, but in general it creates variation in cogging torque. The basic structure of PMSM and design parameters of slot width are shown in Figure 7.6 and Table 7.3 respectively. The simulation has been carried out on Ansys–Maxwell for the variation from 2 mm to 5 mm in width of slot mouth and the results are presented below.

The cogging torque variation under different widths of slot mouth are shown in Figure 7.7 This analysis clearly indicates that the increment of width directly influences on cogging torque as it affect the magnetic conductance of air gap [37]. Without affecting the torque output, the cogging torque can be reduced by changing the conventional design pattern, but it increses the cost of manufactuing. Also toothless stators are lead to zero cogging torque and it's used only in small power machines which are less than 1 Hp.

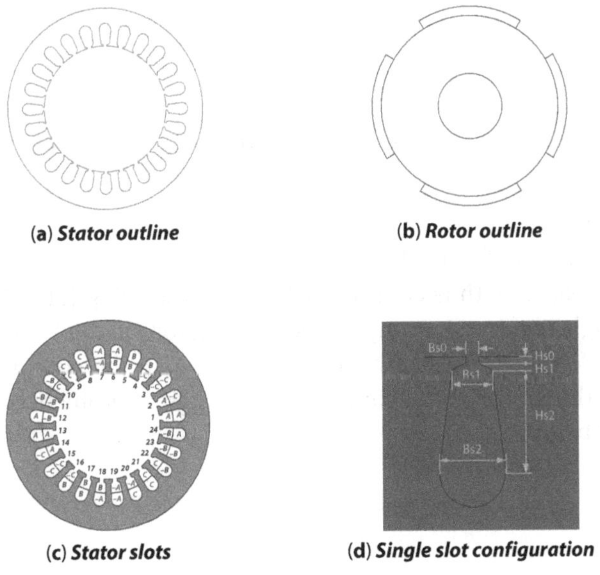

Figure 7.6 Basic structure of PMSM.

Table 7.3 Slot width details.

Parameter	*Value*
Hs0	0.5 mm
Hs1	1 mm
Hs2	8.2
Bs0	2 mm – 5 mm (variation)
Bs1	5.6 mm
Bs2	7.6 mm

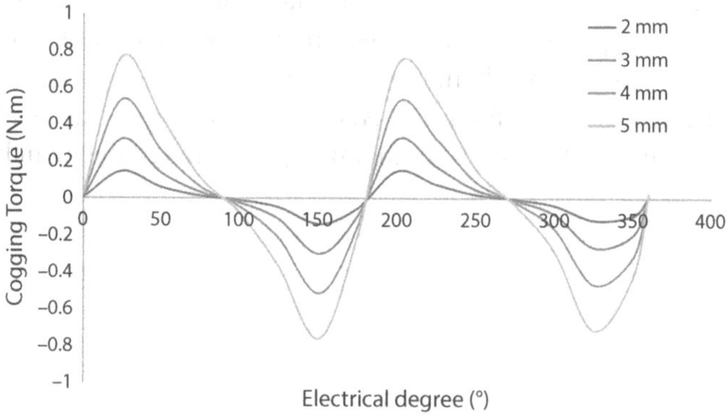

Figure 7.7 Cogging torque variation under different widths of slot mouth.

(iii) Skewing

The phase shifting of stator laminations or changing the position fixture of PMs along axial length is commonly known as skewing [38]. Appropriate skewing angle required to mitigate the cogging torque can be derived from Eqn. (7.1). If we choose skew angle of θ_{sk} to span the complete cogging cycle, then the fundamental component of cogging torque become zero as mentioned below.

$$\theta_{co} = \theta_{sk} = \frac{2\pi}{f_c}$$

$$T_{sk} = \frac{1}{\theta_{sk}} \int_0^{\theta_{sk}} T_C(\theta)\, d\theta$$

$$= \frac{1}{\theta_{sk}} \sum_{m=1}^{\infty} \int_0^{\frac{2\pi}{f_c}} T_m \sin(m f_C \theta + \varphi_m)\, d\theta \tag{7.5}$$

$$= \frac{1}{\theta_{sk}} \sum_{m=1}^{\infty} \left[\frac{-T_m \cos(m f_C \theta + \varphi_m)}{m f_C} \right]_0^{\frac{2\pi}{f_c}} = 0 \tag{7.6}$$

Also this leads to the elimination of entire cogging torque. But due to fringing effects and effects present in the PMSM due to the tolerances of magnets and stator laminations, in practical cases the cogging torque cannot be mitigated fully by skewing itself. Further skewing of rotor magnets is a significant method but it requires very precise manufacturing approach. To avoid such issues, choosing of multiple magnets and step skewing as shown in Figure 7.8 are highly preferred [33].

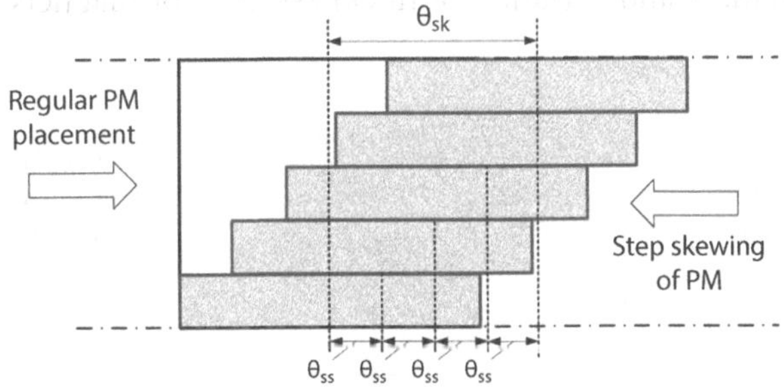

Figure 7.8 Step skewing of PM's.

If N-steps of skewing considered, the cogging torque can be derived [33] as,

$$T_C(\theta) = \sum_{l=1}^{L} \sum_{m=1}^{\infty} T_{ck} \sin\left(ms\left[\theta - (l-1)\frac{2\pi}{f_c} \right] \right) \tag{7.7}$$

The experimental study [15] has proven that, when number of steps increases, the average torque increases so that the torque pulsation decreases. These both methods lead to reduce the flat portion in the induced Electro-Motive Force (EMF) and creating almost sinusoidal nature which is not so desirable when power controllers are employed for the drive, as the input current is rectangular and impacts on cogging torque negatively.

A 4 poles, 24 slots PMSM with data mentioned in Table 7.1 is simulated in Maxwell and the analysis of variation of cogging torque is presented in Figure 7.9. As stated earlier, while we increase the skewing value, the cogging torque is weakened, even to 100 %, i.e zero N.m.

B. Manufacturing Irregularities

Accordance with the regular components which contribute the cogging torque, there are few additional components exist at the manufacturing stage of PMSM, especially when the motors are mass-produced [39]. It includes either misplacement in stator teeth or variation in width and thickness of PM. A detailed study found on effect of variation in PM's assembly, width and thickness in [39], and it summarizes, the misplacement in assembly and variation of thickness of rotor magnets are directly

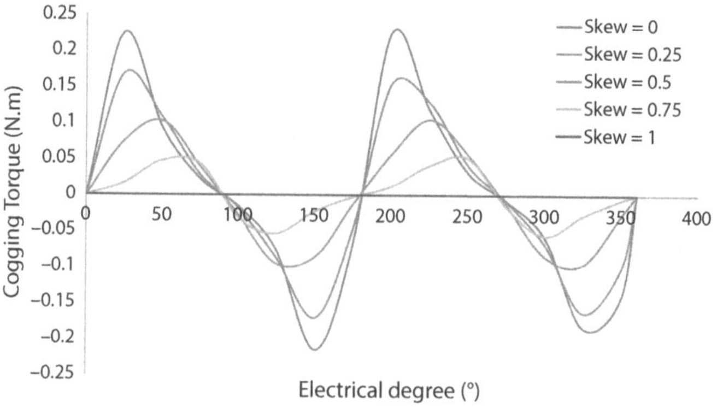

Figure 7.9 Variation of cogging torque under the different skew widths.

proportional to additional components which cause the cogging torque, whereas variation of magnet width does not have any significant change on the same. Further, various parameters influence [33] on the stator lamination isotropy, say punching, stamping, grain orientation, stacking, etc.

These parameters may also affect the cogging torque as they highly change the CPMR. Also the misalignment in eccentricity will lead to variation in permeance value and progressively affects the cogging torque as it alters the characteristic frequency components in the air-gap magnetic field [40]. Consider the basic constructional view of PMSM in Figure 7.10, where stator and rotor positions are indicated.

An analysis proved the rigorous change in the torque output, even for 2 mm of dynamic eccentricity in the alignment of stator and rotor[18]. In order to investigate, the rotor deviation 'β' (16.1°, 17.3° and 18°) has been considered over 'Y' axis as shown in Figure 7.11 and corresponding

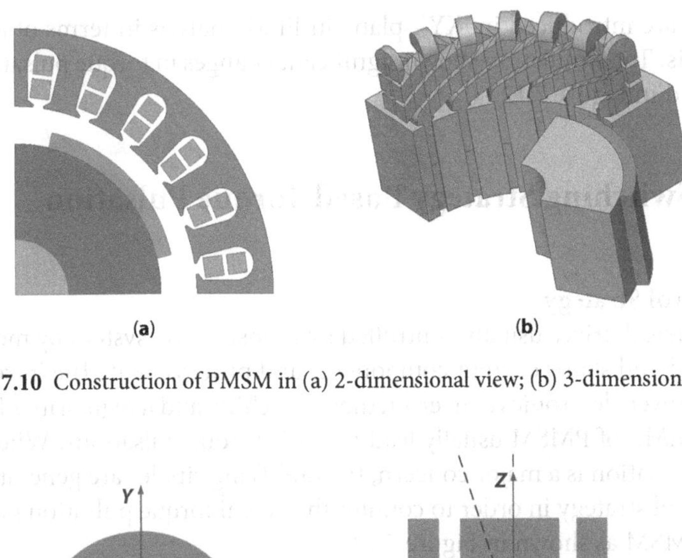

Figure 7.10 Construction of PMSM in (a) 2-dimensional view; (b) 3-dimensional view.

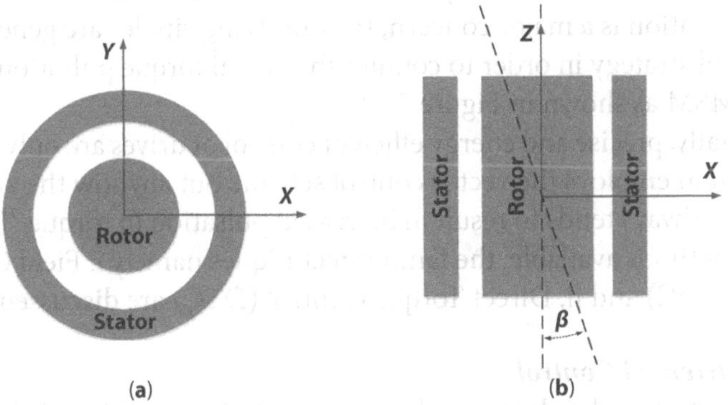

Figure 7.11 Rotor deviation over 'Y' axis in (a) 'XY' plane and (b) 'XZ' plane.

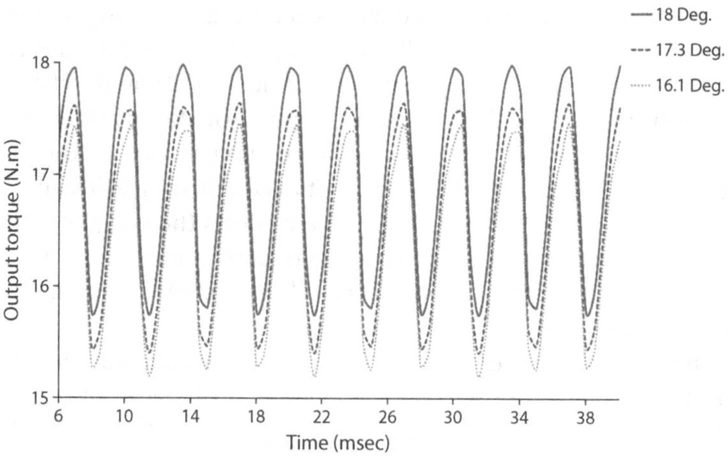

Figure 7.12 Torque response for change in dynamic eccentricity.

changes are introduced in 'XY'- plane in FEM analysis in terms of length in 'X' axis. The simulation shown significant changes in torque pulsation as presented in Figure 7.12.

7.3 Switching Strategy Based Torque Pulsation

A. Control Strategy

The electrical drives usually controlled by a closed loop system by measuring speed and stator current components and processing it. The introduction of power electronics convertor, their switching and irregularities found in back EMF of PMSM usually lead towards torque pulsations. When the torque pulsation is a major concern, the nullifying ripples are generated by the control strategy in order to counter the actual torque pulsation present in the PMSM as shown in Figure 7.13.

Especially, precise and energy efficient control of drives are only possible if the system employs the vector control scheme but anyhow the switching variations always tends to result in increased pulsation in torque. There are various methods available, the famous techniques namely i. Field Oriented Control (FOC) and ii. Direct Torque Control (DTC) are discussed here.

i. Field Oriented Control

The FOC strategy has been used in many rotating machines [41] for last three decades with multiple variation as it always proves the less current

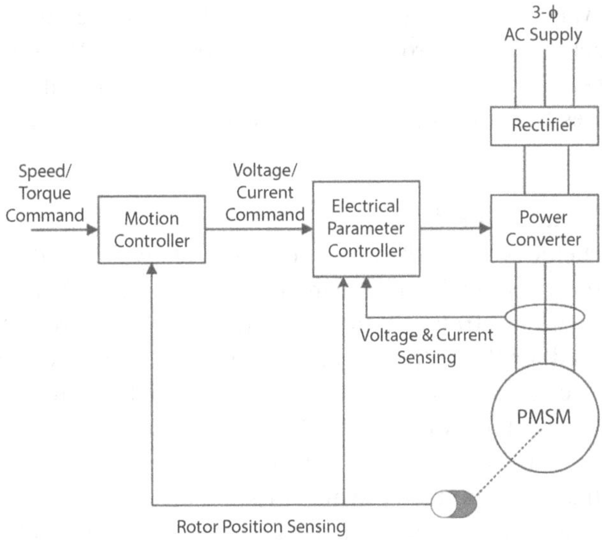

Figure 7.13 Control strategy of PMSM drive.

ripples in dynamic operations. FOC directly control the stator currents which are represented by a vector. Since the three phase current components are time-variant, it's difficult to determine the counter-acting strategy for the particular system. But FOC in general, transforms the time and speed variant current components into time-invariant *d* and *q* components through Clarke and Park transformation as shown in Figure 7.14. So the

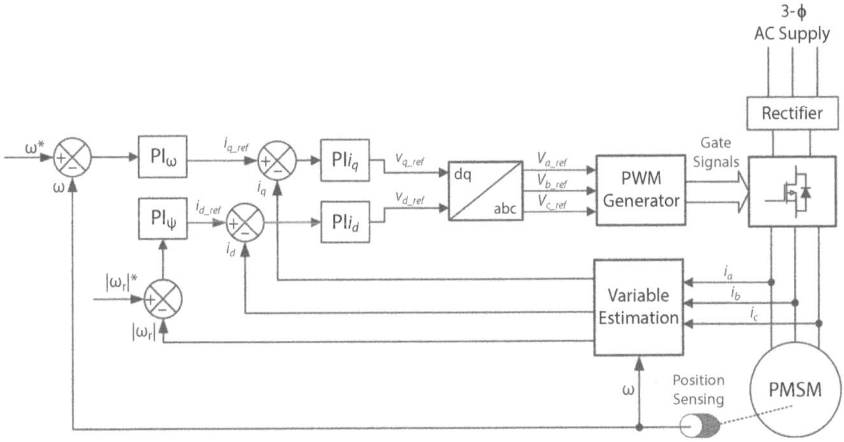

Figure 7.14 Basic scheme of FOC.

other time-variant parameters like speed and torque are not affected and the wide operation of controller is absolutely possible [42]. This configuration is very similar like DC separately excited motor, so controlling is easy, but it requires the knowledge of the rotor position [43] to determine the speed.

ii. Direct Torque Control

The DTC scheme is also well known control strategy [44] and been used as better substitute for FOC [20]. The main motive of this technique is to control the flux and electromagnetic torque directly by choosing the voltage vector required for the Voltage Source Inverter (VSI) from flux and torque errors analytic as shown in Figure 7.15. Unlike the FOC, rotor position is determined from stator variables in DTC. This control scheme proven [20] fast torque responses over steady-state and transient operating conditions.

The evaluation of control strategies are simulated in MATLAB: Simulink to realize the torque pulsation and its variation. The brief model of PMSM and required equations are reconstituted from various detailed studies [20], [33], [42], [43] and presented over here.

The stator and rotor flux linkages in different reference frames are shown in Figure 7.16. The voltage and flux linkage equations and electromagnetic torque equations in the '*dq*' reference frame are generally written as [20],

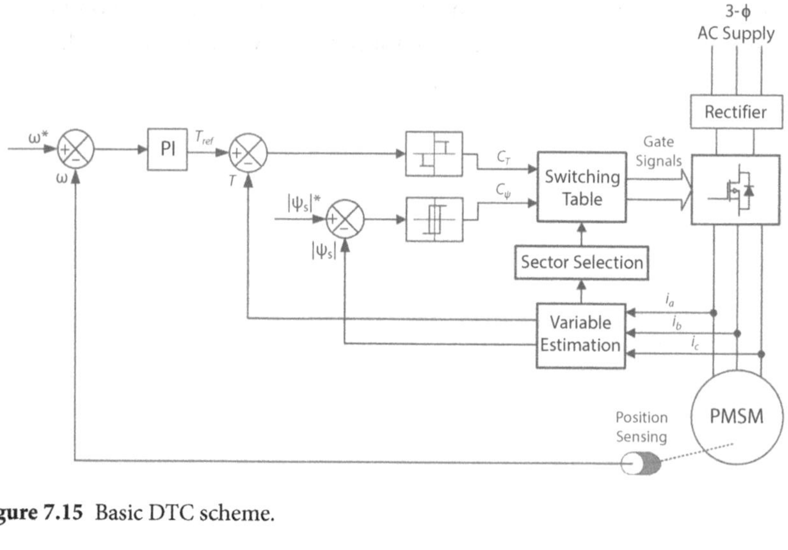

Figure 7.15 Basic DTC scheme.

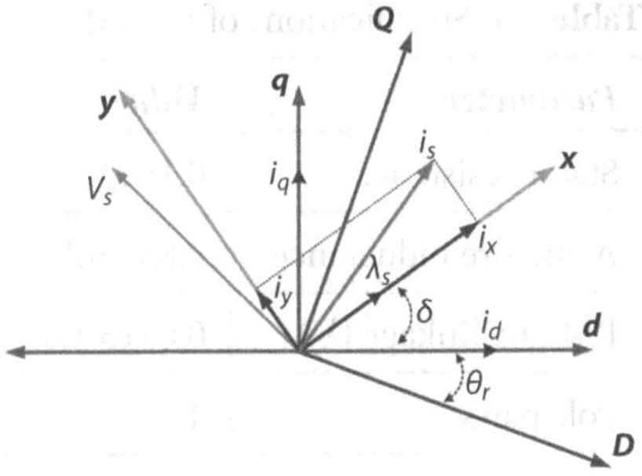

Figure 7.16 The stator and rotor flux linkages in different reference frames.

$$\begin{pmatrix} v_d \\ v_q \end{pmatrix} = \begin{pmatrix} r_s + L_d p & -L_q\omega_r \\ L_d\omega_r & r_s + L_q p \end{pmatrix} \begin{pmatrix} i_d \\ i_q \end{pmatrix} + \lambda_r\omega_r \begin{pmatrix} 0 \\ 1 \end{pmatrix} \tag{7.8}$$

$$\begin{aligned} \lambda_q &= L_q i_q \\ \lambda_d &= L_d i_d + \lambda_r \end{aligned} \tag{7.9}$$

$$T_{em} = \frac{3P|\lambda_s|}{4L_d L_q}\left[2\lambda_r L_q \sin\delta - |\lambda_s|(L_q - L_d)\sin 2\delta\right] \tag{7.10}$$

where, v_d & v_q = Stator voltage
i_d & i_q = Stator current
r_s = Stator resistance
L_d & L_q = Armature synchronous inductances
λ_q & λ_d = Independent '*dq*' axis flux linkages
λ_s & λ_r = Stator and rotor flux linkages
P = Pole Pair
δ = Angle between the stator and the rotor flux linkage

The simulation of FOC is carried out in Matlab – Simulink for the specification mentioned in Table 7.4 and results are shown for the stator current and output torque in Figure 7.17.

Table 7.4 Specifications of PMSM.

Parameter	*Value*
Stator resistance	0.18 Ω
Armature inductance	0.835 mH
PM flux linkage (λ_r)	0.0715 V.s
Pole pairs	4

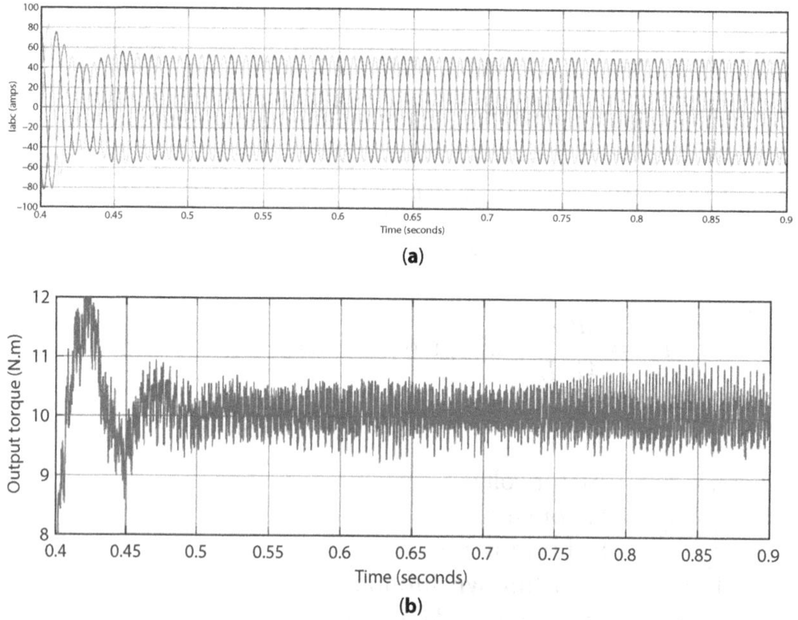

Figure 7.17 Stator current and torque measurement for FOC method.

Further in DTC methodology voltage profiles are chosen from vector plane which is divided into six regions as shown Figure 7.18 and appropriate switching states are chosen from Table 7.5.

The simulation of DTC is carried out in Matlab – Simulink for the specification mentioned in table and results are shown for the stator current and output torque in Figure 7.19. As stated earlier, both methods clearly indicate the significant torque ripple [45].

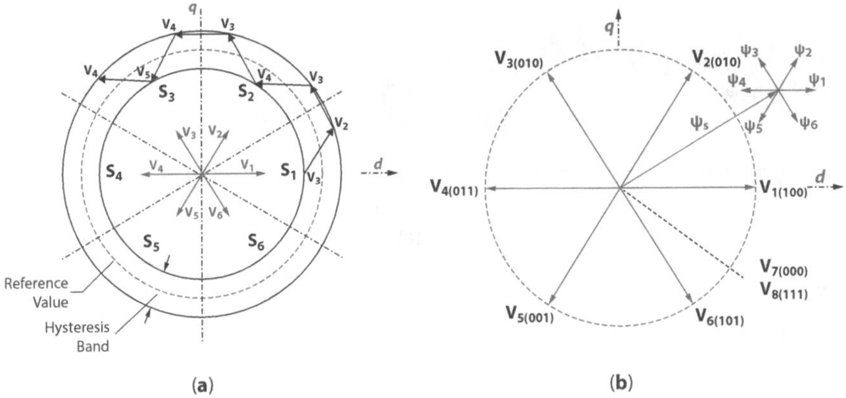

Figure 7.18 (a) Trajectory of stator flux, (b) Voltage profiles and respective stator flux changes.

Table 7.5 DTC switching table.

		Stator flux sectors					
C_ψ	C_T	S_1	S_2	S_3	S_4	S_5	S_6
+1	+1	V_2	V_3	V_4	V_5	V_6	V_1
	0	V_8	V_7	V_8	V_7	V_8	V_7
	-1	V_6	V_1	V_2	V_3	V_4	V_5
-1	+1	V_3	V_4	V_5	V_6	V_1	V_2
	0	V_7	V_8	V_7	V_8	V_7	V_8
	-1	V_5	V_6	V_1	V_2	V_3	V_4

B. Transient

i. Dead time/Commutation delay

The short circuit conditions in VSI–DC bus are usually prevented by introducing a dead-time interval [28] in switching signals to avoid any delays in commutation of power switches.

The 3-ϕ VSI topology and current flow direction in one phase leg are shown in Figure 7.20. Also the depiction of switching pattern with the consideration of dead time is shown in Figure 7.21.

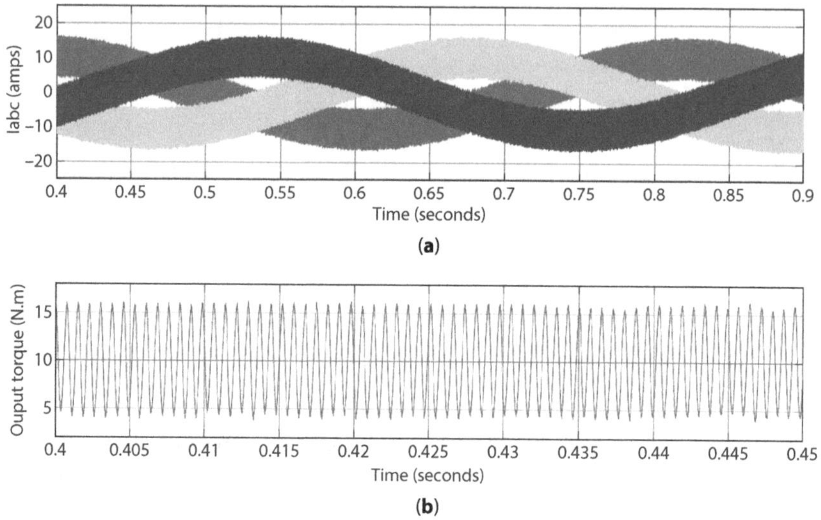

Figure 7.19 Stator current and torque measurement for DTC method.

These delays create unwanted harmonic, especially in the stationary reference frame the fifth and seventh order (or sixth order in synchronous reference frame) harmonics [46]. It cause the distortion in inverter output voltage [47], [48] and increasing the torque pulsation [49]. According to [46] and [49] it's clearly visible that dead-time and commutation are significantly involved in torque pulsation.

ii. Braking

In order to save the energy regenerative braking is mostly preferred than conventional mechanical braking set-up in many of the drives. Also it saves the space in EV's and diminishes the mechanical wear and friction, hence the improvement in flexibility of torque control [50]. The regenerative braking uses the back EMF of the motor, generally reversing current flow towards source. During this current reversal, a commutation torque pulsation [31] is produced because of the time delay between outgoing phase and incoming phase, involving the phase inductance. Especially when PMSM uses sensor based control strategy such FOC or where external sensors are not possible but high precision is important, the commutation torque becomes much dominant [51]. The dead time or commutation delay found in both the scenario has been simulated by altering the carrier wave of the Pulse Width Modulation (PWM) in the electrical parameter controller block shown in Figure 7.13.

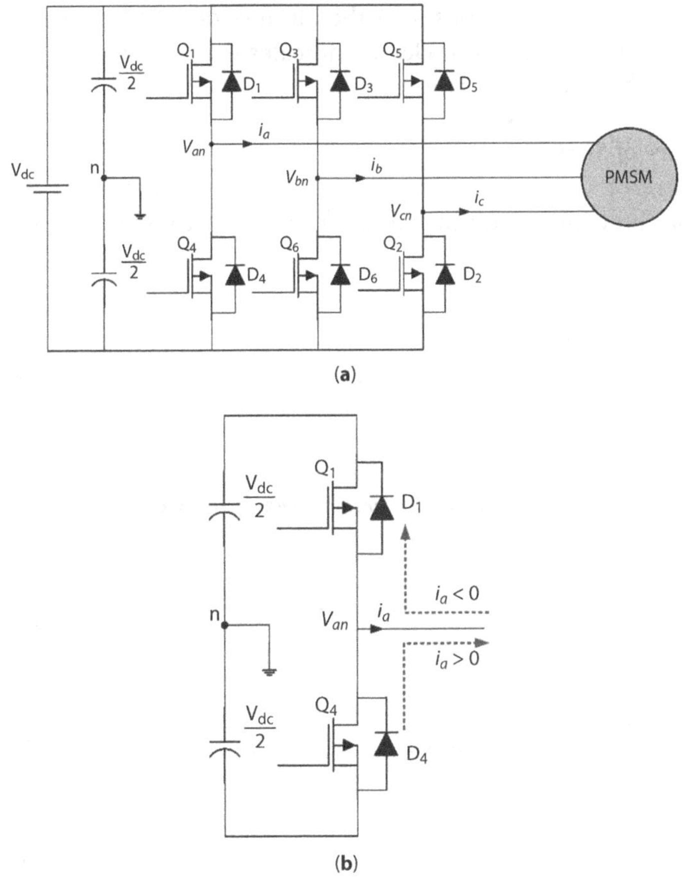

Figure 7.20 (a) 3-ϕ MOSFET based VSI topology (b) Current flow direction.

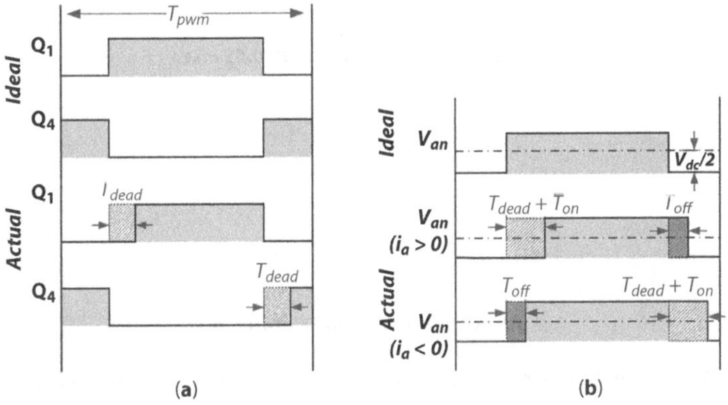

Figure 7.21 Switching pattern with the consideration of dead time.

The output torque responses for the variance of PWM carrier signal are shown in Figure 7.22, and it clearly indicates that the pulsation of torque increases when the delay is more.

iii. Failure of Switch

The failure of switch in inverter circuit is another important parameter which has influence on torque pulsation as well as EV's performance while

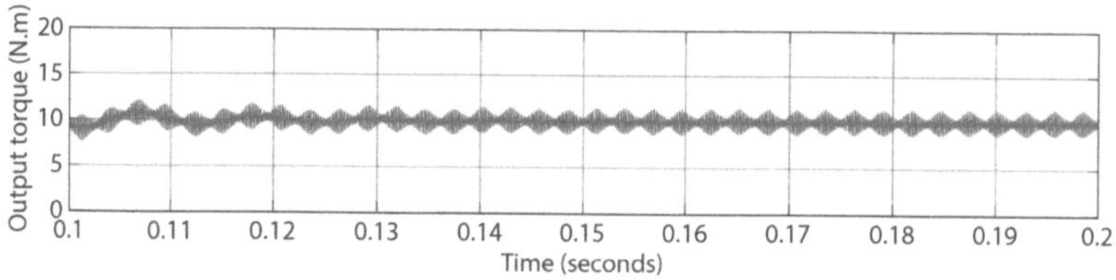

(a) ***PWM carrier signal width 0.24 msec***

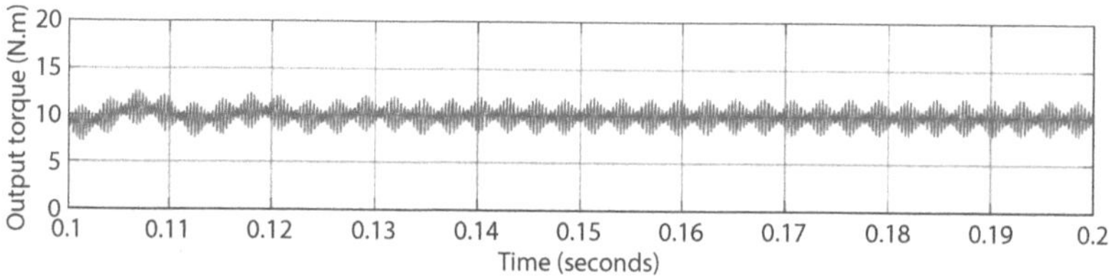

(b) ***PWM carrier signal width 0.33 msec***

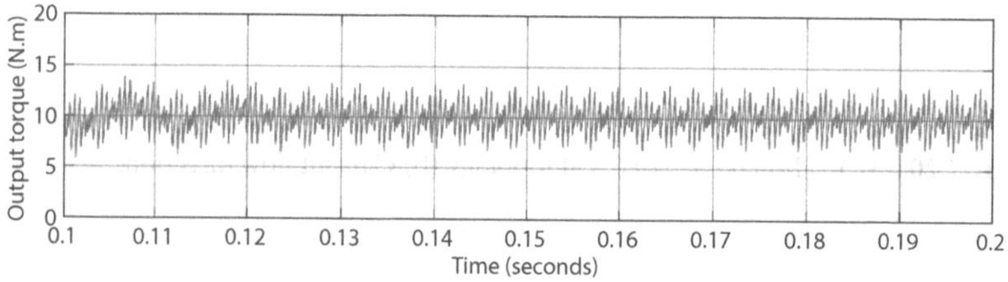

(c) ***PWM carrier signal width 0.55 msec***

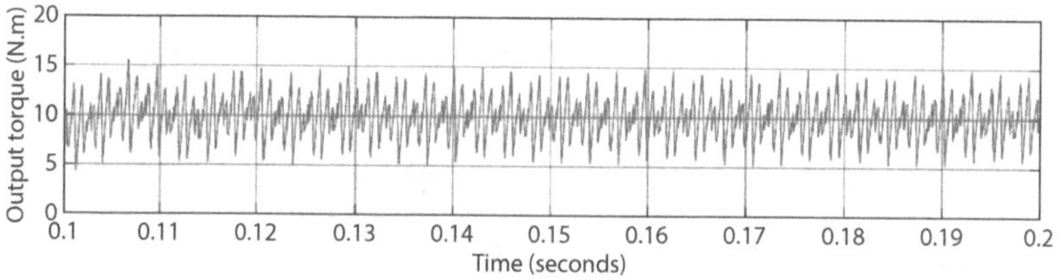

(d) ***PWM carrier signal width 1.6 msec***

Figure 7.22 Output torque responses for different width of PWM carrier signal (or delay).

driving. There are certain failures in VSI like anyone switch/both switches failure in any leg, i.e either open circuit or short circuit, one leg/two legs diode failure, freewheeling diode failure and one leg/two legs open circuit are addressed in [32].

The handling of such faults in PMSM drive is quite difficult [52] by its control strategy and it usually causes high torque pulsation and reduces the average torque. To analyse the effect of switch failure, the switch Q_4 in leg 1 and switches Q_5 and Q_2 in leg 3 as from Figure 7.20 (a) are short circuited one after other and simulated at t=0.5 sec. The variation in stator current and output torque are presented for both the cases in Figure 7.23 and Figure 7.24.

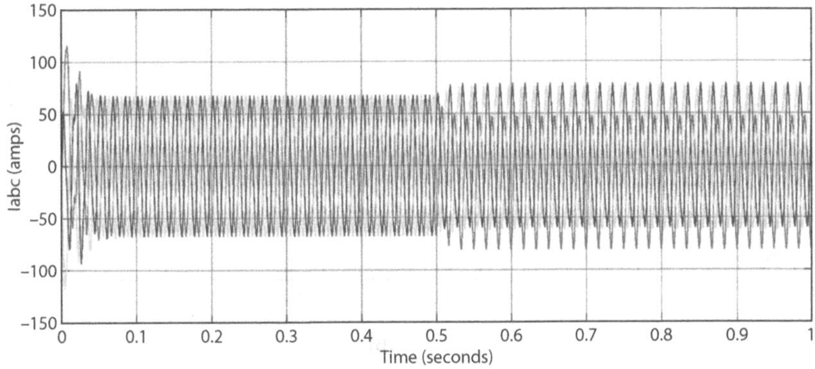

(a)

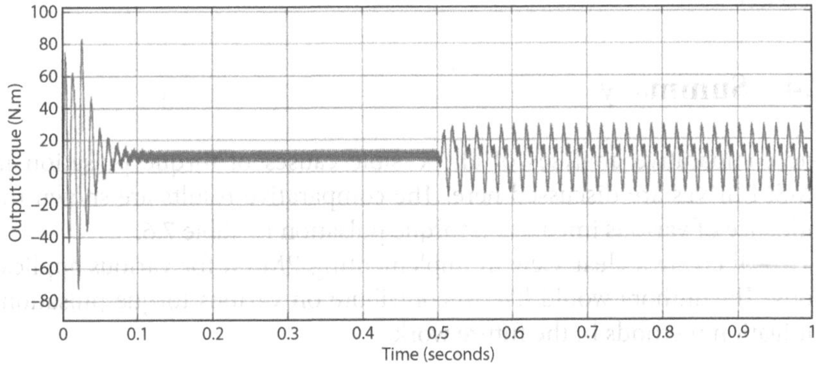

(b)

Figure 7.23 Stator current and torque measurement (switch Q_4 failure at t=0.5 sec).

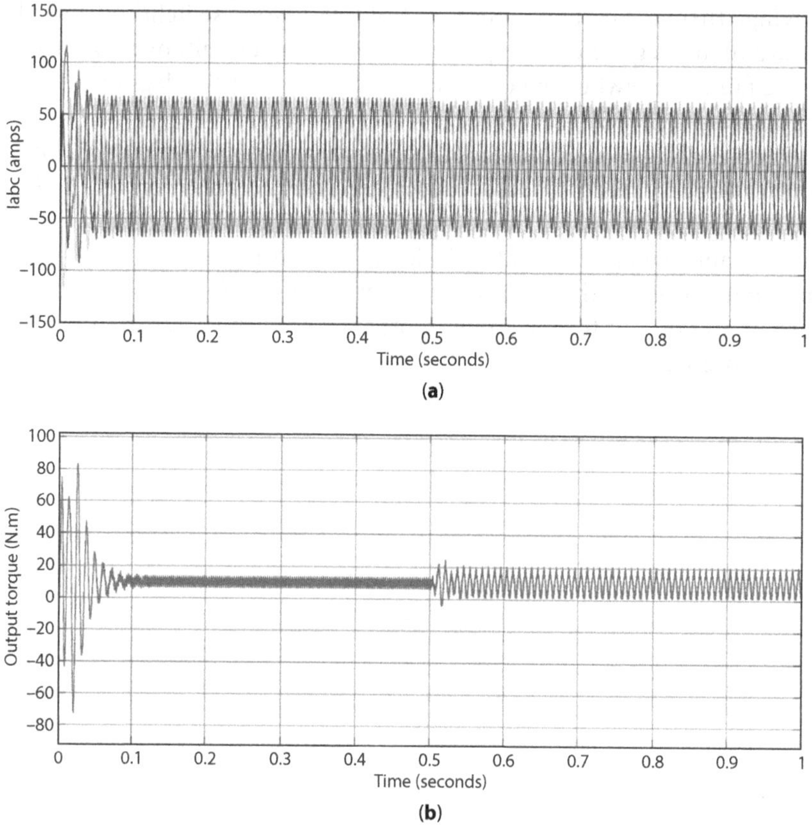

Figure 7.24 Stator current and torque measurement (switch Q_5 and Q_2 failure at t=0.5 sec).

7.4 Summary

The comparative analyses of all possible causes of torque pulsation in PMSM drives are discussed here. The comparative results are shown the influence of various impacts on torque pulsation in Table 7.6.

It will create a clear view in implementing PMSM for various applications. The authors would like to contribute on various torque pulsation-mitigation methods in the future work.

Table 7.6 Comparative analysis.

Type	**Classification**	**Influence**
Cogging Torque	Flux remanance	***Medium***
	Improper slot width	***High***
	Skewing	***High***
	Manufacturing irregularities	***Medium***
Control Strategy	FOC	***High***
	DTC	***High***
Transient	Dead time/commutation delay	***Medium***
	Braking	***Medium***
	Failure of switch	***High***

References

[1] Z. Yang, F. Shang, I. P. Brown, and M. Krishnamurthy, "Comparative study of interior permanent magnet, induction, and switched reluctance motor drives for EV and HEV applications," *IEEE Trans. Transp. Electrif.*, vol. 1, no. 3, pp. 245–254, 2015.

[2] R. Krishnan, "Application Characteristics of Permanent Magnet Synchronous and Brushless dc Motors for Servo Drives," *IEEE Trans. Ind. Appl.*, vol. 27, no. 5, pp. 986–996, 1991.

[3] E. M. Tsampouris, M. E. Beniakar, and A. G. Kladas, "Geometry optimization of PMSMs comparing full and fractional pitch winding configurations for aerospace actuation applications," *IEEE Trans. Magn.*, vol. 48, no. 2, pp. 943–946, 2012.

[4] D. Ying, S. Liwei, and C. Shumei, "Development of PMSM drives for hybrid electric car applications," *IEEE Trans. Magn.*, vol. 43, no. 1, pp. 434–437, 2007.

[5] X. Huang, J. Wang, Y. Xu, Y. Zhu, T. Tang, and M. Lv, "Research on Heave Compensation Control of Floating Crane Based on Permanent Magnet Synchronous Motor," *ICCAIS 2018 - 7th Int. Conf. Control. Autom. Inf. Sci.*, no. Iccais, pp. 331–336, 2018.

[6] A. Emadi, K. Rajashekara, S. S. Williamson, and S. M. Lukic, "Topological overview of hybrid electric and fuel cell vehicular power system architectures

and configurations," *IEEE Trans. Veh. Technol.*, vol. 54, no. 3, pp. 763–770, 2005.

[7] K. M. Rahman and S. E. Schulz, "High performance fully digital switched reluctance motor controller for vehicle propulsion," *Conf. Rec. - IAS Annu. Meet. (IEEE Ind. Appl. Soc.*, vol. 1, no. C, pp. 18–25, 2001.

[8] and R. Z. C. C. Chan, K. T. Chau, J. Z. Jiang, W. Xia, M. Zhu, "Novel permanent magnet motor drives for electric vehicles," *Power Transm. Des.*, vol. 43, no. 2, pp. 331–339, 1996.

[9] S. S. Williamson, S. M. Lukic, and A. Emadi, "Comprehensive drive train efficiency analysis of hybrid electric and fuel cell vehicles based on motor-controller efficiency modeling," *IEEE Trans. Power Electron.*, vol. 21, no. 3, pp. 730–740, 2006.

[10] W. Ren, Q. Xu, Q. Li, and C. Le, "Reducing cogging torque and suppressing torque ripple in PMASynRM for EV/HEV applications," *IEEE Transp. Electrif. Conf. Expo, ITEC Asia-Pacific 2014 - Conf. Proc.*, pp. 1–6, 2014.

[11] Z. Q. Zhu and D. Howe, "Influence of design parameters on cogging torque in permanent magnet machines," *IEEE Trans. Energy Convers.*, vol. 15, no. 4, pp. 407–412, 2000.

[12] D. S. P. Salminen and P. J. Pyrhönen, "TORQUE RIPPLE OF PERMANENT MAGNET MACHINES WITH CONCENTRATED WINDINGS," in *ISEF 2005 - XII International Symposium on Electromagnetic Fields in Mechatronics, Electrical and Electronic Engineering*, 2005, pp. 1–7.

[13] G. H. Jang and D. K. Lieu, "Vibration Reduction In Electric Machine By Magnet Interlacing," *IEEE Trans. Magn.*, vol. 28, no. 5, pp. 3024–3026, 1992.

[14] A. Wang, H. Li, and C. T. Liu, "On the material and temperature impacts of interior permanent magnet machine for electric vehicle applications," *IEEE Trans. Magn.*, vol. 44, no. 11 PART 2, pp. 4329–4332, 2008.

[15] J. A. Güemes, A. M. Iraolagoitia, J. I. Del Hoyo, and P. Fernández, "Torque analysis in permanent-magnet synchronous motors: A comparative study," *IEEE Trans. Energy Convers.*, vol. 26, no. 1, pp. 55–63, 2011.

[16] M. S. Islam, S. Mir, and T. Sebastian, "Issues in reducing the cogging torque of mass-produced permanent-magnet brushless DC motor," *IEEE Trans. Ind. Appl.*, vol. 40, no. 3, pp. 813–820, 2004.

[17] R. Islam, I. Husain, A. Fardoun, and K. McLaughlin, "Permanent Magnet Synchronous Motor Magnet Designs with Skewing for Torque Ripple and Cogging Torque Reduction," *IEEE Trans. Ind. Appl.*, vol. 45, no. 1, pp. 1552–1559, 2008.

[18] M. Thiele and G. Heins, "Computationally Efficient Method for Identifying Manufacturing Induced Rotor and Stator Misalignment in Permanent Magnet Brushless Machines," *IEEE Trans. Ind. Appl.*, vol. 52, no. 4, pp. 3033–3040, 2016.

[19] K. Gulez, A. A. Adam, and H. Pastaci, "Torque ripple and EMI noise minimization in PMSM using active filter topology and field-oriented control," *IEEE Trans. Ind. Electron.*, vol. 55, no. 1, pp. 251–257, 2008.

[20] L. Zhong, M. F. Rahman, W. Y. Hu, and K. W. Lim, "Analysis of direct torque control in permanent magnet synchronous motor drives," *IEEE Trans. Power Electron.*, vol. 12, no. 3, pp. 528–536, 1997.

[21] Y. Zhang and Z. Jianguo, "Direct Torque Control of Permanent Magnet Synchronous Motor With Reduced Torque Ripple and Commutation Frequency," *IEEE Trans. Power Electron.*, vol. 26, no. 1, pp. 235–248, 2011.

[22] H. Zhu, X. Xiao, and Y. Li, "Torque ripple reduction of the torque predictive control scheme for permanent-magnet synchronous motors," *IEEE Trans. Ind. Electron.*, vol. 59, no. 2, pp. 871–877, 2012.

[23] C. Xia, J. Zhao, Y. Yan, and T. Shi, "A novel direct torque control of matrix converter-fed PMSM drives using duty cycle control for torque ripple reduction," *IEEE Trans. Ind. Electron.*, vol. 61, no. 6, pp. 2700–2713, 2014.

[24] D. Casadei, F. Profumo, G. Serra, and A. Tani, "FOC and DTC: Two viable schemes for induction motors torque control," *IEEE Trans. Power Electron.*, vol. 17, no. 5, pp. 779–787, 2002.

[25] G. D. Andreescu, C. I. Pitic, F. Blaabjerg, and I. Boldea, "Combined flux observer with signal injection enhancement for wide speed range sensorless direct torque control of IPMSM drives," *IEEE Trans. Energy Convers.*, vol. 23, no. 2, pp. 393–402, 2008.

[26] G. Foo, S. Sayeef, and M. F. Rahman, "Sensorless direct torque and flux-controlled IPM synchronous motor drive at very low speed without signal injection," *Aust. J. Electr. Electron. Eng.*, vol. 6, no. 3, pp. 221–231, 2009.

[27] Z. Wang, J. Chen, M. Cheng, and K. T. Chau, "Field-oriented control and direct torque control for paralleled VSIs Fed PMSM drives with variable switching frequencies," *IEEE Trans. Power Electron.*, vol. 31, no. 3, pp. 2417–2428, 2016.

[28] Z. Tang and B. Akin, "A New LMS Algorithm Based Deadtime Compensation Method for PMSM FOC Drives," *IEEE Trans. Ind. Appl.*, vol. 54, no. 6, pp. 6472–6484, 2018.

[29] Z. Tang and B. Akin, "Suppression of Dead-Time Distortion Through Revised Repetitive Controller in PMSM Drives," *IEEE Trans. Energy Convers.*, vol. 32, no. 3, pp. 918–930, 2017.

[30] N. Mutoh and S. Member, "Driving and Braking Torque Distribution Methods for Electric Vehicles on Roads With Low Friction Coefficient," *IEEE Trans. Ind. Electron.*, vol. 59, no. 10, pp. 3919–3933, 2012.

[31] Y. Cao, T. Shi, X. Niu, X. Li, and C. Xia, "A Smooth Torque Control Strategy for Brushless DC Motor in Braking Operation," *IEEE Trans. Energy Convers.*, vol. 33, no. 3, pp. 1443–1452, 2018.

[32] O. Gundogmus *et al.*, "Comparison of failure modes, effect analysis and reliability of electric machine drives," *2019 IEEE Int. Electr. Mach. Drives Conf. IEMDC 2019*, pp. 916–921, 2019.

[33] R. Krishnan, *Permanent Magnet Synchronous and Brushless DC Motor Drives*. 2009.

[34] S. A. Saied, K. Abbaszadeh, A. Tenconi, and S. Vaschetto, "New approach to cogging torque simulation using numerical functions," *IEEE Trans. Ind. Appl.*, vol. 50, no. 4, pp. 2420–2426, 2014.
[35] A. M. T. Corp., *Sintered Neodymium-Iron-Boron Magnets.* 770 Linden Avenue, Rochester, NY 14625, 2011.
[36] S. Y. Cho, W. G. Shin, J. S. Park, and W. H. Kim, "A Torque Compensation Control Scheme of PMSM Considering Wide Variation of Permanent Magnet Temperature," *IEEE Trans. Magn.*, vol. 55, no. 2, pp. 1–5, 2019.
[37] Q. Chen, H. Shu, and L. Chen, "Simulation analysis of cogging torque of permanent magnet synchronous motor for electric vehicle," *J. Mech. Sci. Technol.*, vol. 26, no. 12, pp. 4065–4071, 2012.
[38] and D. M. A. Cassat, M. Williams, "Effect of skew on brushless dc motor exciting forces," in *Proceedings of the Annual Symposium on Incremental Motion Control System and Device*, 1993, p. 148.
[39] L. Gašparin, A. Černigoj, S. Markič, and R. Fišer, "Additional cogging torque components in permanent-magnet motors due to manufacturing imperfections," *IEEE Trans. Magn.*, vol. 45, no. 3, pp. 1210–1213, 2009.
[40] A. Rezig, M. R. Mekideche, and A. Djerdir, "Impact of eccentricity and demagnetization faults on magnetic noisegeneration in brushless permanent magnet DC motors," *J. Electr. Eng. Technol.*, vol. 6, no. 3, pp. 356–363, 2011.
[41] F. Blaschke, "The Principle of Field Orientation as Applied to the NEW Transvector Closed-Loop System for Rotating-Field Machines," *Siemens Rev.*, vol. 34, no. 3, pp. 217–220, 1972.
[42] J. Qian and A. Rahman, "Analysis of Field Oriented Control for Permanent Magnet Hysteresis Synchronous Motors," *IEEE Trans. Ind. Appl.*, vol. 29, no. 6, pp. 1156–1163, 1993.
[43] M. G. A. Khlaief, M. Boussak, M. Jemli, "Field-oriented control of interior permanent magnet synchronous motor with a space vector modulation," in *9th International conference on Sciences and Techniques of Automatic control & computer engineering*, 2008.
[44] I. Takahashi and T. Noguchi, "A New Quick-Response and High-Efficiency Control Strategy of an Induction Motor," *IEEE Trans. Ind. Appl.*, vol. IA-22, no. 5, pp. 820–827, 1986.
[45] F. Korkmaz, I. Topaloğlu, M. F. Çakir, and R. Gürbüz, "Comparative performance evaluation of FOC and DTC controlled PMSM drives," *Int. Conf. Power Eng. Energy Electr. Drives*, no. May, pp. 705–708, 2013.
[46] A. Cichowski, J. Nieznanski, and S. Member, "Self-Tuning Dead-Time Compensation Method for Voltage-Source Inverters," vol. 3, no. 2, pp. 72–75, 2005.
[47] O. P. Drives, A. R. Mu, and T. A. Lipo, "On-Line Dead-Time Compensation Technique for," vol. 14, no. 4, pp. 683–689, 1999.
[48] T. Sukegawa, K. Kamiyama, K. Mizuno, T. Matsui, and T. Okuyama, "Fully digital vector-controlled PWM VSI-Fed AC Drives with an Inverter

Dead-Time Compensation Strategy," *IEEE Trans. Ind. Appl.*, vol. 27, no. 3, pp. 552–559, 1991.

[49] N. Urasaki, T. Senjyu, K. Uezato, and T. Funabashi, "Adaptive dead-time compensation strategy for permanent magnet synchronous motor drive," *IEEE Trans. Energy Convers.*, vol. 22, no. 2, pp. 271–280, 2007.

[50] M. J. Yang, H. L. Jhou, B. Y. Ma, and K. Ka. Shyu, "A cost-effective method of electric brake with energy regeneration for electric vehicles," *IEEE Trans. Ind. Electron.*, vol. 56, no. 6, pp. 2203–2212, 2009.

[51] T. D. Batzel and K. Y. Lee, "Commutation torque ripple minimization for permanent magnet synchronous machines with Hall effect position feedback," *IEEE Trans. Energy Convers.*, vol. 13, no. 3, pp. 257–262, 1998.

[52] B. A. Welchko, T. M. Jahns, W. L. Soong, and J. M. Nagashima, "IPM synchronous machine drive response to symmetrical and asymmetrical short circuit faults," *IEEE Trans. Energy Convers.*, vol. 18, no. 2, pp. 291–298, 2003.

Index

Also of Interest

Check out these other related titles from Scrivener Publishing

ELECTRIC VEHICLE DESIGN: Design, Simulation and Applications, Edited by Krishan Arora, Suman Lata Tripathi, and Himashu Sharma, ISBN: 9781394204373. This book will serve as a definitive guide to conceptual and practical knowledge about the design of hybrid electrical vehicles (HEV), battery electrical vehicles (BEV), fuel cell electrical vehicles (FCEV), plug-in hybrid electrical vehicles (PHEV), and efficient EV charging techniques with advanced tools and methodologies for students, engineers, and academics alike.

INDUSTRIAL CONTROL SYSTEMS, Edited by Vipin Chandra Pal, Suman Lata Tripathi, and Souvik Ganguli, ISBN: 9781119829256. This volume serves as a comprehensive guide in the journey of industrial control systems with a multidisciplinary approach to the key engineering problems in the 21st century.

MODERN AUTOMOTIVE ELECTRICAL SYSTEMS: Theory and Applications, Edited by Pedram Asef, Sanjeevikumar Padmanaban, and Andrew Lapthorn, ISBN: 1119801047. Presenting the concepts and advances of modern automotive electrical systems, this volume, written and edited by a global team of experts, also goes into the practical applications for the engineer, student, and other industry professionals.

NANODEVICES FOR INTEGRATED CIRCUIT DESIGN, Edited by Suman Lata Tripathi, Abhishek Kumar, K. Srinivasa Rao, and Prasantha R. Mudimela, ISBN: 9781394185788. Written and edited by a team of experts in the field, this important new volume broadly covers the design of nano-devices and their integrated applications in digital and analog integrated circuits (IC) design.

EXPLAINABLE MACHINE LEARNING MODELS AND ARCHITECTURES: Real-Time System Implementation, Edited by Suman Lata Tripathi and Mufti Mahmud, ISBN: 9781394185849. This cutting-edge new volume covers the hardware architecture implementation, the software implementation approach, and the efficient hardware of machine learning applications.

MACHINE LEARNING TECHNIQUES FOR VLSI CHIP DESIGN, Edited by Abhishek Kumar, Suman Lata Tripathi, and K. Srinivasa Rao, ISBN: 9781119910398. This cutting-edge new volume covers the hardware architecture implementation, the software implementation approach, and the efficient hardware of machine learning applications with FPGA or CMOS circuits, and many other aspects and applications of machine learning techniques for VLSI chip design.

INTELLIGENT GREEN TECHNOLOGIES FOR SMART CITIES, Edited by Suman Lata Tripathi, Souvik Ganguli, Abhishek Kumar, and Tengiz Magradze, ISBN: 9781119816065. Presenting the concepts and fundamentals of smart cities and developing "green" technologies, this volume, written and edited by a global team of experts, also goes into the practical applications that can be utilized across multiple disciplines and industries, for both the engineer and the student.

DESIGN AND DEVELOPMENT OF EFFICIENT ENERGY SYSTEMS, edited by Suman Lata Tripathi, Dushyant Kumar Singh, Sanjeevikumar Padmanaban, and P. Raja, ISBN 9781119761631. Covering the concepts and fundamentals of efficient energy systems, this volume, written and edited by a global team of experts, also goes into the practical applications that can be utilized across multiple industries, for both the engineer and the student.

Electrical and Electronic Devices, Circuits, and Materials: Technical Challenges and Solutions, edited by Suman Lata Tripathi, Parvej Ahmad Alvi, and Umashankar Subramaniam, ISBN 9781119750369. Covering every aspect of the design and improvement needed for solid-state electronic devices and circuit and their reliability issues, this new volume also includes overall system design for all kinds of analog and digital applications and developments in power systems.

Green Energy: Solar Energy, Photovoltaics, and Smart Cities, edited by Suman Lata Tripathi and Sanjeevikumar Padmanaban, ISBN 9781119760764. Covering the concepts and fundamentals of green energy, this volume, written and edited by a global team of experts, also goes into the practical applications that can be utilized across multiple industries, for both the engineer and the student.

HYDROGEN ELECTRICAL VEHICLES, Edited by Mehmet Sankir and Nurdan Demirci Sankir,ISBN: 9781394166381. Hydrogen electrical vehicles are an essential component of the "Green New Deal" and this book covers cutting-edge technologies designed for fuel-cell-powered cars.

PAVEMENT ASSET MANAGEMENT, By Ralph Haas, W.R. Hudson, and Lynne Cowe Falls, ISBN: 9781119038702. Whether you are new to managing pavement assets or a seasoned practitioner, this book provides the most up-to-date information in the field on the history, technology, practice and look ahead.

Printed in the USA/Agawam, MA
September 16, 2026

910661.054